W9-DIR-774

Tarascon Pocket Pharmacopoeia®

2007 Classic Shirt-Pocket Edition

Tarascon Pocket Pharmacopoeia®
2007 Classic Shirt-Pocket Edition

"Desire to take medicines ... distinguishes man from animals." *Sir William Osler*

FOR TARASCON BOOKS/SOFTWARE, VISIT **WWW.TARASCON.COM**

Tarascon Pocket Pharmacopoeia®
- Classic Shirt-Pocket Edition
- Deluxe Labcoat Pocket Edition
- PDA software for Palm OS® or Pocket PC®

Other Tarascon Pocketbooks
- Tarascon Internal Medicine & Critical Care Pocketbook
- Tarascon Pediatric Emergency Pocketbook
- Tarascon Primary Care Pocketbook
- Tarascon Pocket Orthopaedica®
- Tarascon Adult Emergency Pocketbook
- How to be a Truly Excellent Junior Medical Student

See faxable order form on page 160

"It's not how much you know, it's how fast you can find the answer."®

Important Caution – Please Read This!
The information in the *Pocket Pharmacopoeia* is compiled from sources believed to be reliable, and exhaustive efforts have been put forth to make the book as accurate as possible. The *Pocket Pharmacopoeia* is edited by a panel of drug information experts with extensive peer review and input from more than 50 practicing clinicians of multiple specialties. Our goal is to provide health professionals focused, core prescribing information in a convenient, organized, and concise fashion. We include FDA-approved dosing indications and those off-label uses that have a reasonable basis to support their use. *However the accuracy and completeness of this work cannot be guaranteed.* Despite our best efforts this book may contain typographical errors and omissions. The *Pocket Pharmacopoeia* is intended as a quick and convenient reminder of information you have already learned elsewhere. The contents are to be used as a guide only, and health care professionals should use sound clinical judgment and individualize therapy to each specific patient care situation. This book is not meant to be a replacement for training, experience, continuing medical education, studying the latest drug prescribing literature, raw intelligence, good karma, or common sense. This book is sold without warranties of any kind, express or implied, and the publisher and editors disclaim any liability, loss, or damage caused by the contents. *If you do not wish to be bound by the foregoing cautions and conditions, you may return your undamaged and unexpired book to our office for a full refund.* Tarascon Publishing is independent from and has no affiliation with pharmaceutical companies. Although drug companies purchase and distribute our books as promotional items, the Tarascon editorial staff alone determine all book content.

Tarascon Pocket Pharmacopoeia® 2007 Classic Edition

   CY1

HOW TO USE THE TARASCON POCKET PHARMACOPOEIA®

The *Tarascon Pocket Pharmacopoeia* arranges drugs by clinical class with a comprehensive index in the back. Trade names are italicized and capitalized. Drug doses shown in mg/kg are generally intended for children, while fixed doses represent typical adult recommendations. Brackets indicate currently available formulations, although not all pharmacies stock all formulations. The availability of generic, over-the-counter, and scored formulations are mentioned. Codes are as follows:

▶ **METABOLISM & EXCRETION**: L = primarily liver, K = primarily kidney, LK = both, but liver > kidney, KL = both, but kidney > liver, LO = liver & onions

♀ **SAFETY IN PREGNANCY**: A = Safety established using human studies, B = Presumed safety based on animal studies, C = Uncertain safety; no human studies and animal studies show an adverse effect, D = Unsafe - evidence of risk that may in certain clinical circumstances be justifiable, X = Highly unsafe - risk of use outweighs any possible benefit. For drugs which have not been assigned a category: + Generally accepted as safe, ? Safety unknown or controversial, - Generally regarded as unsafe.

▶ **SAFETY IN LACTATION**: + Generally accepted as safe, ? Safety unknown or controversial, - Generally regarded as unsafe. Many of our "+" listings are from the *AAP* policy "The Transfer of Drugs and Other Chemicals Into Human Milk" (see www.aap.org) and may differ from those recommended by the manufacturer.

© **DEA CONTROLLED SUBSTANCES**: I = High abuse potential, no accepted use (eg, heroin, marijuana), II = High abuse potential and severe dependence liability (eg, morphine, codeine, hydromorphone, cocaine, amphetamines, methylphenidate, secobarbital). Some states require triplicates. III = Moderate dependence liability (eg, *Tylenol #3, Vicodin*), IV = Limited dependence liability (benzodiazepines, propoxyphene, phentermine), V = Limited abuse potential (eg, *Lomotil*).

$ **RELATIVE COST**: Cost codes used are "per month" of maintenance therapy (eg, antihypertensives) or "per course" of short-term therapy (eg, antibiotics). Codes are calculated using average wholesale prices (at press time in US dollars) for the most common indication and route of each drug at a typical adult dosage. For maintenance therapy, costs are calculated based upon a 30 day supply or the quantity that might typically be used in a given month. For short-term therapy (ie, 10 days or less), costs are calculated on a single treatment course. When multiple forms are

Code	Cost
$	< $25
$$	$25 to $49
$$$	$50 to $99
$$$$	$100 to $199
$$$$$	≥ $200

available (eg, generics), these codes reflect the least expensive generally available product. When drugs don't neatly fit into the classification scheme above, we have assigned codes based upon the relative cost of other similar drugs. *These codes should be used as a rough guide only*, as (1) they reflect cost, not charges, (2) pricing often varies substantially from location to location and time to time, and (3) HMOs, Medicaid, and buying groups often negotiate quite different pricing. Your mileage may vary. Check with your local pharmacy if you have any question.

🍁 **CANADIAN TRADE NAMES**: Unique common Canadian trade names not used in the US are listed after a maple leaf symbol. Trade names used in both nations or only in the US are displayed without such notation.

PAGE INDEX FOR TABLES*

*Please see the *Deluxe Edition* of the *Tarascon Pocket Pharmacopoeia* for extra large format tables as follows. General: P450 Enzymes. CV: CAD 10-year risk, HTN risk stratification AND treatment. Heme: Warfarin interactions. Ophth: Visual acuity screen.

ABBREVIATIONS IN TEXT

AAP - American Academy of Pediatrics
ac - before meals
ADHD - attention deficit & hyperactivity disorder
AHA - American Heart Association
ANC - absolute neutrophil count
ASA - aspirin
bid - twice per day
BP - blood pressure
BPH - benign prostatic hypertrophy
CAD - coronary artery disease
cap - capsule
CMV - cytomegalovirus
CNS - central nervous system
COPD - chronic obstructive pulmonary disease
CPZ - chlorpromazine
CrCl - creatinine clearance

d - day
D5W - 5% dextrose
DPI - dry powder inhaler
elem - elemental
ET - endotracheal
EPS - extrapyramidal symptoms
g - gram
gtts - drops
GERD - gastroesophageal reflux dz
GU - genitourinary
h - hour
HAART - highly active antiretroviral therapy
HCTZ - hydrochlorothiazide
HRT - hormone replacement therapy
HSV - herpes simplex virus
HTN - hypertension
IM - intramuscular
INR - international normalized ratio
IU - international units

IV - intravenous
JRA - juvenile rheumatoid arthritis
kg - kilogram
LFTs - liver fxn tests
LV - left ventricular
mcg - microgram
MDI - metered dose inhaler
mEq - milliequivalent
mg - milligram
MI - myocardial infarction
min - minute
mL - milliliter
mo - months old
ng - nanogram
NHLBI - National Heart, Lung, and Blood Institute
NS - normal saline
NYHA - New York Heart Association
N/V - nausea/vomiting
OA - osteoarthritis
pc - after meals
PO - by mouth
PR - by rectum

prn - as needed
q - every
qhs - at bedtime
qid - four times/day
qod - every other day
q pm - every evening
RA - rheumatoid arthritis
SC - subcutaneous
soln - solution
supp - suppository
susp - suspension
tab - tablet
TB - tuberculosis
TCAs - tricyclic antidepressants
TIA - transient ischemic attack
tid - 3 times/day
TNF - tumor necrosis factor
tiw - 3 times/week
UC - ulcerative colitis
UTI - urinary tract infection
wk - week
yo - years old

THERAPEUTIC DRUG LEVELS

Drug	Level	Optimal Timing
amikacin peak	20-35 mcg/ml	30 minutes after infusion
amikacin trough	<5 mcg/ml	Just prior to next dose
carbamazepine trough	4-12 mcg/ml	Just prior to next dose
cyclosporine trough	50-300 ng/ml	Just prior to next dose
digoxin	0.8-2.0 ng/ml	Just prior to next dose
ethosuximide trough	40-100 mcg/ml	Just prior to next dose
gentamicin peak	5-10 mcg/ml	30 minutes after infusion
gentamicin trough	<2 mcg/ml	Just prior to next dose
lidocaine	1.5-5 mcg/ml	12-24 hours after start of infusion
lithium trough	0.6-1.2 meq/l	Just prior to first morning dose
NAPA	10-30 mcg/ml	Just prior to next procainamide dose
phenobarbital trough	15-40 mcg/ml	Just prior to next dose
phenytoin trough	10-20 mcg/ml	Just prior to next dose
primidone trough	5-12 mcg/ml	Just prior to next dose
procainamide	4-10 mcg/ml	Just prior to next dose
quinidine	2-5 mcg/ml	Just prior to next dose
theophylline	5-15 mcg/ml	8-12 hrs after once daily dose
tobramycin peak	5-10 mcg/ml	30 minutes after infusion
tobramycin trough	<2 mcg/ml	Just prior to next dose
valproate trough (epilepsy)	50-100 mcg/ml	Just prior to next dose
valproate trough (mania)	45-125 mcg/ml	Just prior to next dose
vancomycin trough	5-10 mcg/ml	Just prior to next dose

PEDIATRIC DRUGS		Age	2m	4m	6m	9m	12m	15m	2y	3y	5y
		Kg	5	6½	8	9	10	11	13	15	19
		Lbs	11	15	17	20	22	24	28	33	42
med	***strength***	***freq***	***teaspoons of liquid per dose** (1 tsp= 5 ml)*								
Tylenol (mg)		q4h	80	80	120	120	160	160	200	240	280
Tylenol (tsp)	160/t	q4h	½	½	¾	¾	1	1	1¼	1½	1¾
ibuprofen (mg)		q6h	-	-	75†	75†	100	100	125	150	175
ibuprofen (tsp)	100/t	q6h	-	-	¾†	¾†	1	1	1¼	1½	1¾
amoxicillin or	125/t	bid	1	1¼	1½	1¾	1¾	2	2¼	2¾	3½
Augmentin	200/t	bid	½	¾	1	1	1¼	1¼	1½	1¾	2¼
(not otitis media)	250/t	bid	½	½	¾	¾	1	1	1¼	1¼	1¾
	400/t	bid	¼	½	½	½	¾	¾	¾	1	1
amoxicillin,	200/t	bid	1	1¼	1¾	2	2	2¼	2¾	3	4
(otitis media)‡	250/t	bid	¾	1¼	1½	1½	1¾	1¾	2¼	2½	3¼
	400/t	bid	½	¾	¾	1	1	1¼	1½	1½	2
Augmentin ES‡	600/t	bid	⅜	½	½	¾	¾	¾	1	1¼	1½
azithromycin*§	100/t	qd	¼†	½†	½	½	½	½	¾	¾	1
(5-day Rx)	200/t	qd	--	¼†	¼	¼	¼	¼	½	½	½
Bactrim/Septra	---	bid	½	¾	1	1	1	1¼	1½	1½	2
cefaclor*	125/t	bid	1	1	1¼	1½	1½	1¾	2	2½	3
"	250/t	bid	½	½	¾	¾	¾	1	1	1¼	1½
cefadroxil	125/t	bid	½	¾	1	1	1¼	1¼	1½	1¾	2¼
"	250/t	bid	¼	½	½	½	¾	¾	¾	1	1
cefdinir	125/t	qd	--	¾†	1	1	1	1¼	1½	1¾	2
cefixime	100/t	qd	½	½	¾	¾	¾	1	1	1¼	1½
cefprozil*	125/t	bid	--	¾†	1	1	1¼	1½	1½	2	2¼
"	250/t	bid	--	½†	½	½	¾	¾	¾	1	1¼
cefuroxime	125/t	bid	--	¾	¾	1	1	1	1½	1¾	2¼
cephalexin	125/t	qid	--	½	¾	¾	1	1	1¼	1½	1¾
"	250/t	qid	--	¼	¼	½	½	½	¾	¾	1
clarithromycin	125/t	bid	½†	½†	½	½	¾	¾	¾	1	1¼
"	250/t	bid	--	--	--	¼	½	½	½	½	¾
dicloxacillin	62½/t	qid	½	¾	1	1	1¼	1¼	1½	1¾	2
loracarbef*	100/t	bid	--	1†	1¼	1½	1½	1¾	2	2¼	3
nitrofurantoin	25/t	qid	¼	½	½	½	½	¾	¾	¾	1
Pediazole	---	tid	½	½	¾	¾	1	1	1	1¼	1½
penicillin V**	250/t	bid-tid	--	1	1	1	1	1	1	1	1
cetirizine	5/t	qd	-	-	½	½	½	½	½	½	½
Benadryl	12.5/t	q6h	½	½	¾	¾	1	1	1¼	1½	2
prednisolone	15/t	qd	¼	½	½	¾	¾	¾	1	1	1¼
prednisone	5/t	qd	1	1¼	1½	1¾	2	2¼	2½	3	3¾
Robitussin	---	q4h	-	-	¼†	¼†	½	½	¾	¾	1
Tylenol w/ codeine		q4h	-	-	-	-	-	-	-	1	1

* Dose shown is for otitis media only; see dosing in text for alternative indications.

† Dosing at this age/weight not recommended by manufacturer.

‡ AAP now recommends high dose (80-90 mg/kg/d) for all otitis media in children; with Augmentin used as ES only.

§Give a double dose of azithromycin the first day.

**AHA dosing for streptococcal pharyngitis. Treat for 10 days.

PEDIATRIC VITAL SIGNS AND INTRAVENOUS DRUGS

Age		Pre-matr	New born	2m	4m	6m	9m	12m	15m	2y	3y	5y
Weight	(Kg)	2	3½	5	6½	8	9	10	11	13	15	19
	(Lbs)	4½	7½	11	15	17	20	22	24	28	33	42
Maint fluids	(ml/h)	8	14	20	26	32	36	40	42	46	50	58
ET tube	(mm)	2½	3/3½	3½	3½	3½	4	4	4½	4½	4½	5
Defib	(Joules)	4	7	10	13	16	18	20	22	26	30	38
Systolic BP	(high)	70	80	85	90	95	100	103	104	106	109	114
	(low)	40	60	70	70	70	70	70	70	75	75	80
Pulse rate	(high)	145	145	180	180	180	160	160	160	150	150	135
	(low)	100	100	110	110	110	100	100	100	90	90	65
Resp rate	(high)	60	60	50	50	50	46	46	30	30	25	25
	(low)	35	30	30	30	24	24	20	20	20	20	20
adenosine	(mg)	0.2	0.3	0.5	0.6	0.8	0.9	1	1.1	1.3	1.5	1.9
atropine	(mg)	0.1	0.1	0.1	0.13	0.16	0.18	0.2	0.22	0.26	0.30	0.38
Benadryl	(mg)	-	-	5	6½	8	9	10	11	13	15	19
bicarbonate	(meq)	2	3½	5	6½	8	9	10	11	13	15	19
dextrose	(g)	1	2	5	6½	8	9	10	11	13	15	19
epinephrine	(mg)	.02	.04	.05	.07	.08	.09	0.1	0.11	0.13	0.15	0.19
lidocaine	(mg)	2	3½	5	6½	8	9	10	11	13	15	19
morphine	(mg)	0.2	0.3	0.5	0.6	0.8	0.9	1	1.1	1.3	1.5	1.9
mannitol	(g)	2	3½	5	6½	8	9	10	11	13	15	19
naloxone	(mg)	.02	.04	.05	.07	.08	.09	0.1	0.11	0.13	0.15	0.19
diazepam	(mg)	0.6	1	1.5	2	2.5	2.7	3	3.3	3.9	4.5	5
fosphenytoin*	(PE)	40	70	100	130	160	180	200	220	260	300	380
lorazepam	(mg)	0.1	0.2	0.3	0.35	0.4	0.5	0.5	0.6	0.7	0.8	1.0
phenobarb	(mg)	30	60	75	100	125	125	150	175	200	225	275
phenytoin*	(mg)	40	70	100	130	160	180	200	220	260	300	380
ampicillin	(mg)	100	175	250	325	400	450	500	550	650	750	1000
ceftriaxone	(mg)	-	-	250	325	400	450	500	550	650	750	1000
cefotaxime	(mg)	100	175	250	325	400	450	500	550	650	750	1000
gentamicin	(mg)	5	8	12	16	20	22	25	27	32	37	47

*Loading doses; fosphenytoin dosed in "phenytoin equivalents".

If you obtained your *Pocket Pharmacopoeia* from a bookstore, please send your address to info@tarascon.com. This allows you to be the first to hear of updates! (We don't sell or distribute our mailing lists, by the way.) The cover woodcut is *The Apothecary* by Jost Amman, Frankfurt, 1574. Several of you knew that "tribbles" are cute furry little alien creatures that are gravid at birth and fortunately yelp in the presence of intruders – in this case menacing Klingons from the original Star Trek TV series. If you've missed this hilarious episode you've got to rent it! We will send a free copy of next year's edition to the first 25 who know what apparition dotty will send us sequestered in a potty to a scaly stare so haughty.

CONVERSIONS	Liquid:	Weight:
Temperature:	1 fluid ounce = 30ml	1 kilogram = 2.2 lbs
F = (1.8) C + 32	1 teaspoon = 5ml	1 ounce = 30 g
C = (F - 32)/(1.8)	1 tablespoon = 15ml	1 grain = 65 mg

Alveolar-arterial oxygen gradient = A-a = 148 - 1.2(PaCO2) - PaO2
[normal = 10-20 mmHg, breathing room air at sea level]

Calculated osmolality = 2Na + glucose/18 + BUN/2.8 + ethanol/4.6
[norm 280-295 meq/L. Na in meq/L; all others in mg/dL]

Pediatric IV maintenance fluids (see table on page 7)
4 ml/kg/hr **or** 100 ml/kg/day for first 10 kg, plus
2 ml/kg/hr **or** 50 ml/kg/day for second 10 kg, plus
1 ml/kg/hr **or** 20 ml/kg/day for all further kg

$$mcg/kg/min = \frac{16.7 \times \text{drug conc [mg/ml]} \times \text{infusion rate [ml/h]}}{\text{weight [kg]}}$$

$$\text{Infusion rate [ml/h]} = \frac{\text{desired mcg/kg/min} \times \text{weight [kg]} \times 60}{\text{drug concentration [mcg/ml]}}$$

Fractional excretion of sodium = $\left[\dfrac{\text{urine Na / plasma Na}}{\text{urine creat / plasma creat}}\right] \times 100\%$
[Pre-renal, etc <1%; ATN, etc >1%]

Anion gap = Na – (Cl + HCO3)   [normal = 10-14 meq/L]

$$\text{Creatinine clearance} = \frac{(\text{lean kg})(140 - \text{age})(0.85 \text{ if female})}{(72)(\text{stable creatinine [mg/dL]})}$$
[normal >80]

Glomerular filtration rate using MDRD equation (ml/min/1.73 m^2)
= 186 x (creatinine)$^{-1.154}$ x (age)$^{-0.203}$ x (0.742 if ♀) x (1.210 if African American)

Body surface area (BSA) = square root of: $\left[\dfrac{\text{height (cm)} \times \text{weight (kg)}}{3600}\right]$
[in m^2]

DRUG THERAPY REFERENCE WEBSITES (selected)

Professional societies or governmental agencies with drug therapy guidelines		
AHRQ	Agency for Healthcare Research and Quality	www.ahcpr.gov
AAP	American Academy of Pediatrics	www.aap.org
ACC	American College of Cardiology	www.acc.org
ACCP	American College of Chest Physicians	www.chestnet.org
ACCP	American College of Clinical Pharmacy	www.accp.com
AHA	American Heart Association	www.americanheart.org
ADA	American Diabetes Association	www.diabetes.org
AMA	American Medical Association	www.ama-assn.org
ATS	American Thoracic Society	www.thoracic.org
ASHP	Amer. Society Health-Systems Pharmacists	www.ashp.org
CDC	Centers for Disease Control & Prevention	www.cdc.gov
CDC	CDC bioterrorism and radiation exposures	www.bt.cdc.gov
IDSA	Infectious Diseases Society of America	www.idsociety.org
MHA	Malignant Hyperthermia Association	www.mhaus.org
NHLBI	National Heart, Lung, & Blood Institute	www.nhlbi.nih.gov
Other therapy reference sites		
Cochrane library		www.cochrane.org
Emergency Contraception Website		www.not-2-late.com
Immunization Action Coalition		www.immunize.org
Int'l Registry for Drug-Induced Arrhythmias		www.qtdrugs.org
Managing Contraception		www.managingcontraception.com
Nephrology Pharmacy Associates		www.nephrologypharmacy.com

ANALGESICS

Antirheumatic Agents - Immunomodulators - TNF Inhibitors
etanercept (**Enbrel**): RA, psoriatic arthritis, ankylosing spondylitis: 50 mg SC q wk. Plaque psoriasis: 50 mg SC twice weekly x 3 months, then 50 mg SC q week. JRA 4-17 yo: 0.8 mg/kg SC q week, to max single dose of 50 mg. Max dose per injection site is 25 mg. [Carton contains 4 dose trays and single-use prefilled syringes. Each dose tray contains one 25 mg single-use vial, one 50 mg/mL syringe, one plunger, and two alcohol swabs.] ▶Serum ♀B ▶- $$$$$
infliximab (**Remicade**): RA: 3 mg/kg IV in combo with methotrexate at 0, 2, and 6 wks. Ankylosing spondylitis: 5 mg/kg IV at 0, 2, and 6 wks. Psoriatic arthritis: 5 mg/kg IV at 0, 2, and 6 wks with or without methotrexate. ▶Serum ♀B ▶? $$$$$

Antirheumatic Agents - Immunomodulators - Other
anakinra (**Kineret**): RA: 100 mg SC daily. [Trade only: 100 mg pre-filled glass syringes with needles, 7 or 28 per box.] ▶K ♀B ▶? $$$$$
leflunomide (**Arava**): RA: 100 mg PO daily x 3 days. Maintenance: 10-20 mg PO daily. [Generic/Trade: Tabs 10, 20 mg. Trade: Tab 100 mg.] ▶LK ♀X ▶- $$$$$

Antirheumatic Agents - Other
azathioprine (**Azasan, Imuran, ✦Immunoprin, Oprisine**): RA: Initial dose 1 mg/kg (50-100 mg) PO daily or divided bid. Increase after 6-8 weeks. [Generic/Trade: Tabs 50 mg, scored. Trade only (Azasan): 75, 100 mg, scored.] ▶LK ♀D ▶- $$$
hydroxychloroquine (**Plaquenil**): RA: start 400-600 mg PO daily, then taper to 200-400 mg daily. SLE: 400 PO daily-bid to start, then taper to 200-400 mg daily. [Generic/Trade: Tabs 200 mg, scored.] ▶K ♀C ▶+ $$
methotrexate (**Rheumatrex, Trexall**): RA, psoriasis: Start with 7.5 mg/week PO single dose or 2.5 mg PO q12h x 3 doses given as a course once weekly. Max dose 20 mg/week. Supplement with 1 mg/day of folic acid. [Trade only (Trexall): Tabs 5, 7.5, 10, 15 mg. Dose Pak (Rheumatrex) 2.5mg (#8,12,16,20,24's). Generic/Trade: Tabs 2.5 mg, scored.] ▶LK ♀X ▶- $$
sulfasalazine (**Azulfidine, Azulfidine EN-tabs, ✦Salazopyrin, Salazopyrin EN, S.A.S.**): RA: 500 mg PO daily-bid after meals up to 1g PO bid. May turn body fluids, contact lenses or skin orange-yellow. [Generic/Trade: Tabs 500 mg, scored. Enteric coated, Delayed-release (EN-Tabs) 500 mg.] ▶L ♀B ▶- $$

Muscle Relaxants
baclofen (**Lioresal, Kemstro**): Spasticity related to MS or spinal cord disease/injury: Start 5 mg PO tid, then increase by 5 mg/dose q 3 days until 20 mg PO tid. Max dose 20 mg qid. [Generic only: Tabs 10, 20 mg. Trade only: (Kemstro) Tabs-orally disintegrating 10, 20 mg.] ▶K ♀C ▶+ $$
carisoprodol (**Soma**): Acute musculoskeletal pain: 350 mg PO tid-qid. Abuse potential. [Generic/Trade: Tabs 350 mg.] ▶LK ♀? ▶- $
chlorzoxazone (**Parafon Forte DSC**): Musculoskeletal pain: 500-750 mg PO tid-qid to start. Decrease to 250 mg tid-qid. [Generic/Trade: Tabs & caplets 250 & 500 mg (Parafon Forte DSC 500 mg tablets scored).] ▶LK ♀C ▶? $
cyclobenzaprine (**Flexeril**): Musculoskeletal pain: Start 5-10 mg PO tid, max 60 mg/day. Not recommended in elderly. [Generic/Trade: Tab 5, 10 mg. Generic only: Tab 7.5 mg.] ▶LK ♀B ▶? $
dantrolene (**Dantrium**): Chronic spasticity related to spinal cord injury, stroke, cerebral palsy, MS: 25 mg PO daily to start, up to max of 100 mg bid-qid if necessary.

Malignant hyperthermia: 2.5 mg/kg rapid IV push q 5-10 minutes continuing until symptoms subside or to a maximum 10 mg/kg/dose. [Generic/Trade: Caps 25, 50, 100 mg.] ▶LK ♀C ▶- $$$$

diazepam (**Valium, Diastat**, ✚**Vivol, E Pam**): Skeletal muscle spasm, spasticity related to cerebral palsy, paraplegia, athetosis, stiff man syndrome: 2-10 mg PO/ PR tid-qid. [Generic/Trade: Tabs 2, 5, 10 mg, trade scored. Generic: Oral solution 5 mg/5 mL & concentrated soln 5 mg/mL. Trade only: (Diastat) Rectal gel: 2.5, 10, 20 mg, in twin packs.] ▶LK ♀D ▶- ©IV $

metaxalone (**Skelaxin**): Musculoskeletal pain: 800 mg PO tid-qid. [Trade only: Tabs 800 mg, scored.] ▶LK ♀? ▶? $$$

methocarbamol (**Robaxin, Robaxin-750**): Acute musculoskeletal pain: 1500 mg PO qid or 1000 mg IM/IV tid x 48-72h. Maintenance: 1000 mg PO qid, 750 mg PO q4h, or 1500 mg PO tid. Tetanus: specialized dosing. [Generic/Trade: Tabs 500 & 750 mg. OTC in Canada.] ▶LK ♀C ▶? $$

orphenadrine (**Norflex**): Musculoskeletal pain: 100 mg PO bid. 60 mg IV/IM bid. [Generic only: 100 mg extended release. OTC in Canada.] ▶LK ♀C ▶? $$

quinine sulfate: Commonly prescribed for nocturnal leg cramps, but not FDA approved for this indication: 260-325 mg PO qhs. Cinchonism with overdose. Hemolysis with G6PD deficiency, hypersensitivity. [Generic: Caps 200, 324, 325 mg. Tabs 260 mg.] ▶L ♀X ▶+ $$

tizanidine (**Zanaflex**): Muscle spasticity due to MS or spinal cord injury: 4-8 mg PO q6-8h prn, max 36 mg/d. [Generic: Tabs 2, 4 mg, scored. Trade only: Caps 2, 4, 6 mg.] ▶LK ♀C ▶? $$$$

Non-Opioid Analgesic Combinations

Ascriptin (aspirin + aluminum hydroxide + magnesium hydroxide + calcium carbonate, Aspir-Mox): Multiple strengths. 1-2 tabs PO q4h. [OTC: Trade only: Tabs 325 mg ASA/50 mg magnesium hydroxide/50 mg aluminum hydroxide/50 mg calcium carb (Ascriptin & Aspir-Mox). 500 mg ASA/33 mg mag hydroxide/33 mg aluminum hydroxide/237 mg calcium carb (Ascriptin Extra Strength).] ▶K ♀D ▶? $

Bufferin (aspirin + calcium carbonate + magnesium oxide + magnesium carbonate): 1-2 tabs/caplets PO q4h. [OTC: Trade only: Tabs/caplets 325 mg ASA/158 mg Ca carbonate/63 mg of Mg oxide/34 mg of M carbonate. Bufferin ES: 500 mg ASA/222.3 mg Ca carb/88.9 mg of Mg oxide/55.6 mg of Mg carb] ▶K ♀D ▶? $

Esgic (acetaminophen + butalbital + caffeine): 1-2 tabs or caps PO q4h. Max 6 in 24 hours. [Generic/Trade: Tabs/caps Esgic is 325/50/40 mg of acetaminophen/ butalbital/caffeine. Esgic Plus is 500/50/40 mg.] ▶LK ♀C ▶? $

Excedrin Migraine (acetaminophen + aspirin + caffeine): 2 tabs/caps/geltabs PO q6h while symptoms persist. Max 8 in 24 hours. [OTC/Generic/Trade: Tabs/caplets/geltabs acetaminophen 250 mg/ASA 250 mg/caffeine 65 mg.] ▶LK ♀D ▶? $

Fioricet (acetaminophen + butalbital + caffeine): 1-2 tabs PO q4h. Max 6 in 24 hours. [Generic/Trade: Tab 325 mg acetaminophen//50 mg butalbital/40 mg caffeine.] ▶LK ♀C ▶? $

Fiorinal (aspirin + butalbital + caffeine, ✚**Tecnal, Trianal**): 1-2 tabs PO q4h. Max 6 tabs in 24 hours. [Generic/Trade: Cap 325 mg aspirin/50 mg butalbital/40 mg caffeine.] ▶KL ♀D ▶- ©III $

Goody's Extra Strength Headache Powder (acetaminophen + aspirin + caffeine): 1 powder PO followed with liquid, or stir powder into a glass of water or other liquid. Repeat in 4-6 hours prn. Max 4 powders/24 hours. [OTC trade only: 260 mg acetaminophen/520 mg ASA/32.5 mg caffeine per powder paper.] ▶LK ♀D ▶? $

Norgesic (orphenadrine + aspirin + caffeine): Multiple strengths; write specific product on Rx. *Norgesic:* 1-2 tabs PO tid-qid. *Norgesic Forte,* 1 tab PO tid-qid. [Generic/Trade: Tabs *Norgesic* 25/385/30 mg of orphenadrine/aspirin/caffeine. *Norgesic Forte* 50/770/60 mg.] ▶KL ♀D ▶? $$

Phrenilin (acetaminophen + butalbital): Tension or muscle contraction headache: 1-2 tabs PO q4h. Max 6 in 24 hours. [Generic/Trade: Caps *Phrenilin* 325/50 mg of acetaminophen/butalbital. Caps *Phrenilin Forte* 650/50 mg.] ▶LK ♀C ▶? $

Sedapap (acetaminophen + butalbital): 1-2 tabs PO q4h. Max 6 tabs in 24 hours. [Generic/Trade: Tab 650 mg acetaminophen//50 mg butalbital.] ▶LK ♀C ▶? $

Soma Compound (carisoprodol + aspirin): 1-2 tabs PO qid. Abuse potential. [Generic/Trade: Tab 200 mg carisoprodol/325 mg ASA.] ▶LK ♀D ▶- $$$

Ultracet (tramadol + acetaminophen): Acute pain: 2 tabs PO q4-6h prn, max 8 tabs/ day for ≤5 days. Adjust dose in elderly & renal dysfunction. Avoid in opioid-dependent patients. Seizures may occur if concurrent antidepressants or seizure disorder. [Generic/Trade: Tab 37.5 tramadol/325 mg acet.] ▶KL ♀C ▶- $$

Nonsteroidal Anti-Inflammatories - COX-2 Inhibitors

celecoxib (*Celebrex*): OA, ankylosing spondylitis: 200 mg PO daily or 100 mg PO bid. RA: 100-200 mg PO bid. Familial adenomatous polyposis: 400 mg PO bid with food. Acute pain, dysmenorrhea: 400 mg x 1, then 200 bid prn. An additional 200 mg dose may be given on day 1 if needed. Contraindicated in sulfonamide allergy. [Trade only: Caps 100, 200, 400 mg.] ▶L ♀C (D in 3rd trimester) ▶? $$$$$

Nonsteroidal Anti-Inflammatories - Salicylic Acid Derivatives

aspirin (*Ecotrin, Bayer, Anacin, ASA, ♥Asaphen, Entrophen, Novasen*): 325-650 mg PO/PR q4-6h. [OTC: Tabs 81, 162, 325, 500, 650 mg. Suppositories 60,120, 125, 200, 300, 600 mg. Rx only: 800 mg controlled-release tabs & 975 mg enteric coated.] ▶K ♀D (D in 3rd trimester) ▶? $

choline magnesium trisalicylate (*Trilisate*): 1500 mg PO bid. [Generic/Trade: Tabs 750. Generic only: Tabs 500, 1000 mg. Solution 500 mg/5mL.] ▶K ♀C (D in 3rd trimester) ▶? $$

diflunisal (*Dolobid*): Pain: 500-1000 mg initially, then 250-500 mg PO q8-12h. RA/OA: 500 mg-1g PO divided bid. [Generic/Trade: Tabs 250 & 500 mg.] ▶K ♀C (D in 3rd trimester) ▶- $$$

salsalate (*Salflex, Disalcid, Amigesic*): 3000 mg/day PO divided q8-12h. [Generic/Trade: Tabs 500 & 750 mg, scored.] ▶K ♀C (D in 3rd trimester) ▶? $$

Nonsteroidal Anti-Inflammatories - Other

Arthrotec (diclofenac + misoprostol): OA: one 50/200 tab PO tid. RA: one 50/200 tab PO tid-qid. If intolerant, may use 50/200 or 75/200 PO bid. Misoprostol is an abortifacient. [Trade only: Tabs 50 mg/200 mcg & 75 mg/200 mcg diclofenac/misoprostol.] ▶LK ♀X ▶- $$$$$

diclofenac (*Voltaren, Voltaren XR, Cataflam, ♥Voltaren Rapide*): Multiple strengths; write specific product on Rx. Immediate or delayed release 50 mg PO bid-tid or 75 mg PO bid. Extended release (Voltaren XR): 100-200 mg PO daily. [Generic/Trade: Tabs, immediate-release (Cataflam) 50 mg, extended-release (Voltaren XR) 100 mg. Generic only: Tabs, delayed-release 25, 50, 75 mg.] ▶L ♀B (D in 3rd trimester) ▶- $$$

etodolac (*Lodine, ♥Ultradol*): Multiple strengths; write specific product on Rx. Immediate release 200-400 mg PO bid-tid. Extended release: 400-1200 mg PO daily. [Generic/Trade: Caps 300 mg. Generic only: Tabs 400, 500 mg. Caps, 200 mg. Tabs, extended-release 400, 500, 600 mg.] ▸L ♀C (D in 3rd trimester) ▸- $$$

flurbiprofen (**Ansaid, ✦Froben, Froben SR**): 200-300 mg/day PO divided bid-qid. [Generic/Trade: Tabs 50, 100 mg.] ▸L ♀B (D in 3rd trimester) ▸+ $$$

ibuprofen (**Motrin, Advil, Nuprin, Rufen**): 200-800 mg PO tid-qid. Peds >6 mo: 5-10 mg/kg PO q6-8h. [OTC: Cap/Liqui-Gel Cap 200 mg. Tabs 100, 200 mg. Chewable tabs 50, 100 mg. Liquid & suspension 50 mg/1.25ml, 100 mg/5 mL, suspension 100 mg/2.5 mL. Infant drops 50 mg/1.25 mL (calibrated dropper). Rx only: Tabs 300, 400, 600, 800 mg.] ▸L ♀B (D in 3rd trimester) ▸+ $

indomethacin (**Indocin, Indocin SR, ✦Rhodacine**): Multiple strengths; write specific product on Rx. Immediate release: 25-50 mg cap PO tid. Sustained release: 75 mg cap PO daily-bid. [Generic/Trade: Caps, immediate-release 25 & 50 mg. Oral suspension 25 mg/5 mL. Cap, sustained-release 75 mg.] ▸L ♀B (D in 3rd trimester) ▸+ $

ketoprofen (**Orudis, Orudis KT, Actron, Oruvail, ✦Rhodis, Rhodis EC, Rhodis SR, Rhovail, Orudis SR**): Immediate release: 25-75 mg PO tid-qid. Extended release: 100-200 mg cap PO daily. [OTC: Tab, immed-release, 12.5 mg. Rx: Generic/Trade: Caps, ext'd-release 200 mg. Generic only: Caps, immed-release 25, 50, 75 mg. Caps, ext'd-release 100,150 mg.] ▸L ♀B (D in 3rd trimester) ▸- $$$

ketorolac (**Toradol**): Moderately severe acute pain: 15-30 mg IV/IM q6h or 10 mg PO q4-6h prn. Combined duration IV/IM and PO is not to exceed 5 days. [Generic/Trade: Tab 10 mg.] ▸L ♀C (D in 3rd trimester) ▸+ $

mefenamic acid (**Ponstel, ✦Ponstan**): Mild to moderate pain, primary dysmenorrhea: 500 mg PO initially, then 250 mg PO q6h prn for ≤1 week. [Trade only: Cap 250 mg.] ▸L ♀D ▸- $$$

meloxicam (**Mobic, ✦Mobicox**): RA/OA: 7.5 mg PO daily. JRA, ≥2 yo: 0.125 mg/kg PO daily. [Generic/Trade: Tabs 7.5, 15 mg. Suspension 7.5 mg/5 mL (1.5 mg/mL).] ▸L ♀C (D in 3rd trimester) ▸? $$$$

nabumetone (**Relafen**): RA/OA: Initial: two 500 mg tabs (1000 mg) daily. May increase to 1500-2000 mg PO daily or divided bid. [Generic/Trade: Tabs 500 & 750 mg.] ▸L ♀C (D in 3rd trimester) ▸- $$$

naproxen (**Naprosyn, Aleve, Anaprox, EC-Naprosyn, Naprelan**): Immediate release: 250-500 mg PO bid. Delayed release: 375-500 mg PO bid (do not crush or chew). Controlled release: 750-1000 mg PO daily. JRA ≤13 kg: 2.5 mL PO bid. 14-25 kg: 5 mL PO bid. 26-38 kg: 7.5 mL PO bid. 500 mg naproxen = 550 mg naproxen sodium. [OTC: Generic/Trade: Tabs 220 mg. OTC Trade: Capsules & Gelcaps immediate-release 200 mg. Rx: Generic/Trade: Tabs immediate-release 250, 375, 500 mg. Delayed release 375, 500 mg. Tabs delayed-release enteric coated (EC-Naprosyn) 375, 500 mg. Tabs, controlled-release (Naprelan) 375, 500 mg. Generic/Trade: Suspension 125 mg/5 mL.] ▸L ♀B (D in 3rd trimester) ▸+ $$$

oxaprozin (**Daypro**): 1200 mg PO daily. [Generic/Trade: Caplets, tabs 600 mg, trade scored.] ▸L ♀C (D in 3rd trimester) ▸- $$$

piroxicam (**Feldene, Fexicam**): 20 mg PO daily. [Generic/Trade: Caps 10 & 20 mg.] ▸L ♀B (D in 3rd trimester) ▸+ $$$

NSAIDs – If one class fails, consider another. *Salicylic acid derivatives*: aspirin, diflunisal, salsalate, Trilisate. *Propionic acids*: flurbiprofen, ibuprofen, ketoprofen, naproxen, oxaprozin. *Acetic acids*: diclofenac, etodolac, indomethacin, ketorolac, nabumetone, sulindac, tolmetin. *Fenamates*: meclofenamate. *Oxicams*: meloxicam, piroxicam. *COX-2 inhibitors*: celecoxib.

sulindac (**Clinoril**): 150-200 mg PO bid. [Generic/Trade: Tabs 150 & 200 mg.] ▶L ♀B (D in 3rd trimester) ▶- $$$

tiaprofenic acid (✿**Surgam, Surgam SR**): Canada only. 600 mg PO daily of sustained release, or 300 mg PO bid of regular release. [Generic/Trade: Tab 300 mg. Trade only: Cap, sustained-release 300 mg. Generic only: Tab 200 mg.] ▶K ♀C (D in 3rd trimester) ▶- $$$

tolmetin (**Tolectin**): 200-600 mg PO tid. [Generic/Trade: Tabs 200 (trade scored) & 600 mg. Cap 400 mg.] ▶L ♀C (D in 3rd trimester) ▶+ $$$

Opioid Agonist-Antagonists

buprenorphine (**Buprenex**): 0.3-0.6 mg IV/IM q6h prn. ▶L ♀C ▶- ©III $

butorphanol (**Stadol, Stadol NS**): 0.5-2 mg IV or 1-4 mg IM q3-4h prn. Nasal spray (Stadol NS): 1 spray (1 mg) in 1 nostril q3-4h. Abuse potential. [Generic only: Nasal spray 1 mg/spray, 2.5 mL bottle (14-15 doses/bottle).] ▶LK ♀C ▶+ ©IV $$$

nalbuphine (**Nubain**): 10-20 mg IV/IM/SC q3-6h prn. ▶LK ♀? ▶? $

pentazocine (**Talwin, Talwin NX**): 30 mg IV/IM q3-4h prn (Talwin). 1 tab PO q3-4h. (Talwin NX = 50 mg pentazocine/0.5 mg naloxone). [Generic/Trade: Tab 50 mg with 0.5 mg naloxone, trade scored.] ▶LK ♀C ▶? ©IV $$$

Opioid Agonists

> **NOTE:** May cause drowsiness and/or sedation, which may be enhanced by alcohol & other CNS depressants. Patients with chronic pain may require more frequent & higher dosing. All opioids are pregnancy class D if used for prolonged periods or in high doses at term.

codeine: 0.5-1 mg/kg up to 15-60 mg PO/IM/IV/SC q4-6h. Do not use IV in children. [Generic: Tabs 15, 30, & 60 mg. Oral soln: 15 mg/5 mL.] ▶LK ♀C ▶? ©II $$

fentanyl (**Duragesic, Actiq, Sublimaze, IONSYS**): Transdermal (Duragesic): 1 patch q72 hrs (some chronic pain pts may req q48 h dosing). May wear >1 patch to achieve the correct analgesic effect. Transmucosal lozenge (Actiq) for breakthrough cancer pain: 200-1600 mcg, goal is 4 lozenges on a stick/day in conjunction with long-acting opioid. Adult analgesia/procedural sedation: 50-100 mcg slow IV over 1-2 minutes; carefully titrate to effect. Analgesia: 50-100 mcg IM q1-2h prn. [Generic/Trade: Transdermal patches 25, 50, 75, 100 mcg/h. Trade only: (Actiq) lozenges on a stick, raspberry flavored 200, 400, 600, 800, 1,200, 1,600 mcg. Trade only: TD patch (Duragesic) 12.5 mcg/h. Trade: IONSYS: Iontophoretic transdermal system: 40 mcg fentanyl per activation; max 6 doses per hour. Max per system is eighty 40 mcg doses over 24 hours.] ▶L ♀C ▶+ ©II $$$$$

hydromorphone (**Dilaudid, Dilaudid-5, ✿Hydromorph Contin**): Adults: 2-4 mg PO q4-6h. Titrate dose as high as necessary to relieve cancer pain or other types of non-malignant pain where chronic opioids are necessary. 0.5-2 mg IM/SC or slow IV q4-6h. 3 mg PR q6-8h. Peds ≤12 yo: 0.03-0.08 mg/kg PO q4-6h prn. 0.015 mg/kg/dose IV q4-6h prn. [Generic only: Tabs 2, 4. Generic/Trade: Tabs 8 mg (8mg trade scored). Liquid 5 mg/5 mL.] ▶L ♀C ▶? ©II $$

FENTANYL TRANSDERMAL DOSE (based on ongoing morphine requirement)*

morphine (IV/IM)	morphine (PO)	Transdermal fentanyl
8-22 mg/day	45-134 mg/day	25 mcg/hr
23-37 mg/day	135-224 mg/day	50 mcg/hr
38-52 mg/day	225-314 mg/day	75 mcg/hr
53-67 mg/day	315-404 mg/day	100 mcg/hr

*For higher morphine doses see product insert for transdermal fentanyl equivalencies.

OPIOIDS*	Approximate equianalgesic		Recommended starting dose			
			Adults >50kg		Children/Adults 8 to 50 kg	
	IV / SC / IM	PO	IV / SC / IM	PO	IV / SC / IM	PO
Opioid Agonists						
morphine	10 mg q3-4h	†30 mg q3-4h / †60 mg q3-4h	10 mg q3-4h	30 mg q3-4h	0.1 mg/kg q3-4h	0.3 mg/kg q3-4h
codeine	75 mg q3-4h	130 mg q3-4h	60 mg q2h	60 mg q3-4h	n/r	1 mg/kg q3-4h
fentanyl	0.1 mg q1h	n/a	0.1 mg q1h	n/a	n/r	n/a
hydromorphone	1.5 mg q3-4h	7.5 mg q3-4h	1.5 mg q3-4h	6 mg q3-4h	0.015 mg/kg q3-4h	0.06 mg/kg q3-4h
hydrocodone	n/a	30 mg q3-4h	n/a	10 mg q3-4h	n/a	0.2 mg/kg q3-4h
levorphanol	2 mg q6-8h	4 mg q6-8h	2 mg q6-8h	4 mg q6-8h	0.02 mg/kg q6-8h	0.04 mg/kg q6-8h
meperidine§	100 mg q3h	300 mg q2-3h	100 mg q3h	n/r	0.75 mg/kg q2-3h	n/r
oxycodone	n/a	30 mg q3-4h	n/a	10 mg q3-4h	n/a	0.2 mg/kg q3-4h
oxymorphone	1 mg q3-4h	n/a	1 mg q3-4h	n/a	n/r	n/r
Opioid Agonist-Antagonist and Partial Agonist						
buprenorphine	0.3-0.4 mg q6-8h	n/a	0.4 mg q6-8h	n/a	0.004 mg/kg q6-8h	n/a
butorphanol	2 mg q3-4h	n/a	2 mg q3-4h	n/a	n/r	n/a
nalbuphine	10 mg q3-4h	n/a	10 mg q3-4h	n/a	0.1 mg/kg q3-4h	n/a
pentazocine	60 mg q3-4h	150 mg q3-4h	n/r	50 mg q4-6h	n/r	n/r

*Approximate dosing, adapted from 1992 AHCPR guidelines, www.ahcpr.gov. IV doses should be titrated slowly with appropriate monitoring. All PO dosing is with immediate-release preparations. Use lower doses initially in those not currently taking opioids. Individualize all dosing, especially in the elderly, children, and patients with chronic pain, opioid tolerance, or hepatic/renal insufficiency. Many recommend initially using lower than equivalent doses when switching between different opioids. Not available = "n/a". Not recommended = "n/r". Methadone is excluded due to poor consensus on equivalence.
†30 mg with around the clock dosing, and 60 mg with a single dose or short-term dosing (ie, the opioid-naive).
§Doses should be limited to <600 mg/24 hrs and total duration of use <48 hrs; not for chronic pain.

levorphanol (*Levo-Dromoran*): 2 mg PO q6-8h prn. [Generic: Tabs 2 mg, scored.] ▶L ♀C ▶? ©II $$$

meperidine (*Demerol*, pethidine): 1-1.8 mg/kg up to 150 mg IM/SC/PO or slow IV q3-4h. 75 mg meperidine IV,IM,SC = 300 mg PO. [Generic/Trade: Tabs 50 (trade scored) & 100 mg. Syrup 50 mg/5 mL.] ▶LK ♀C ▶+ ▶+ ©II $$$ *[handwritten: 12.5 mg IV Shivering]*

methadone (*Dolophine, Methadose*, ♥*Metadol*): Severe pain in opioid-tolerant patients: 2.5-10 mg IM/SC/PO q3-4h. Titrate dose as high as necessary to relieve cancer pain or other types of non-malignant pain where chronic opioids are necessary. Opioid dependence: 20-100 mg PO daily. Treatment >3 wks is maintenance and only permitted in approved treatment programs. [Generic/Trade: Tabs 5, 10 mg. Dispersible tabs 40 mg. Oral concentrate: 10 mg/mL. Generic only: Oral soln 5 & 10 mg/5 mL.] ▶L ♀C ▶– ©II $

morphine (*MS Contin, Kadian, Avinza, Roxanol, Oramorph SR, MSIR, DepoDur*, ♥*Statex, M-Eslon, M.O.S., Doloral*): Controlled-release tabs (*MS Contin, Oramorph SR*): Start at 30 mg PO q8-12h. Controlled-release caps (*Kadian*): 20 mg PO q12-24h. Extended-release caps (*Avinza*): Start at 30 mg PO daily. Do not break, chew, or crush *MS Contin* or *Oramorph SR*. *Kadian* & *Avinza* caps may be opened & sprinkled in applesauce for easier administration; however the pellets should not be crushed or chewed. 0.1-0.2 mg/kg up to 15 mg IM/SC or slow IV q4h. Titrate dose as high as necessary to relieve cancer pain or other types of non-malignant pain where chronic opioids are necessary. [Generic/Trade: Tabs, immediate-release: 15 & 30 mg. Trade only: Caps 15 & 30 mg. Generic/Trade: Oral soln: 10 mg/5 mL, 10 mg/2.5 mL, 20 mg/5 mL, 20 mg/mL (concentrate) & 100 mg/5 mL (concentrate). Rectal suppositories 5, 10, 20 & 30 mg. Controlled-release tabs (*MS Contin, Oramorph SR*) 15, 30, 60, 100; 200 mg *MS Contin* only. Controlled-release caps (*Kadian*) 20, 30, 50, 60 & 100 mg. Extended release caps (*Avinza*) 30, 60, 90 & 120 mg.] ▶LK ♀C ▶+ ©II $$$$

oxycodone (*Roxicodone, OxyContin, Percolone, OxyIR, OxyFAST*, ♥*Endocodone, Supeudol*): Immediate-release preparations: 5 mg PO q4-6h prn. Controlled-release (*OxyContin*): 10-40 mg PO q12h (No supporting data for shorter dosing intervals for controlled-release tabs.) Titrate dose as high as necessary to relieve cancer pain or other types of non-malignant pain where chronic opioids are necessary. Do not break, chew, or crush controlled release preparations. [Generic/Trade: Immediate-release: Tabs (scored) & caps 5 mg. Tabs 15, 30 mg. Oral soln 5 mg/5 mL. Oral concentrate 20 mg/mL. Controlled-release tabs: 10, 20, 40, 80 mg. Generic only: Immediate release tabs 10,20 mg.] ▶L ♀C ▶– ©II $$$$

oxymorphone (*Numorphan, Opana*): 10-20 mg PO q4-6h (immediate release) or 5 mg q12h (extended release) in opioid-naïve, 1h pre- or 2h post meals. 5 mg PR q4-6h, prn. 1-1.5 mg IM/SC q4-6h prn. 0.5 mg IV q4-6h prn, increase dose until pain adequately controlled. [Trade: Ext'd release tabs (*Opana ER*) 5,10,20,40 mg. Immed. release (*Opana IR*) 5,10 mg. Supp (*Numorphan*) 5 mg.] ▶L ♀C ▶? ©II $

propoxyphene (*Darvon-N, Darvon Pulvules*): 65-100 mg PO q4h prn. [Generic/Trade: Caps 65 mg. Trade only: Tabs 100 mg (Darvon-N).] ▶L ♀C ▶+ ©IV $

Opioid Analgesic Combinations

NOTE: Refer to individual components for further information. May cause drowsiness and/or sedation, which may be enhanced by alcohol & other CNS depressants. Opioids, carisoprodol, and butalbital may be habit-forming. Avoid exceeding 4 g/day of acetaminophen in combination products. Caution people who drink ≥3 alcoholic drinks/day to limit acetaminophen use to 2.5 g/day due to additive liver toxicity. Opioids commonly cause constipation -- concurrent laxatives are recommended. All opioids are pregnancy class D if used for prolonged periods or in high doses at term.

Anexsia (hydrocodone + acetaminophen): Multiple strengths; write specific product on Rx. 1 tab PO q4-6h prn. [Generic/Trade: Tabs 5/325, 5/500, 7.5/325, 7.5/650 mg hydrocodone/mg acetaminophen, scored.] ▶LK ♀C ▶– ©III $

Capital with Codeine suspension (acetaminophen + codeine): 15 mL PO q4h prn. >12 yo use adult dose. 7-12 yo 10 mL/dose q4-6h prn. 3-6 yo 5 mL/dose q4-6h prn. [Generic = oral soln. Trade = suspension. Both codeine 12 mg and acetaminophen 120 mg per 5 mL (trade, fruit punch flavor).] ▶LK ♀C ▶? ©V $

Combunox (oxycodone + ibuprofen): 1 tab PO q6h prn for ≤7 days. Max 4 tabs/24h. [Trade only: Tab 5 mg oxycodone/400 mg ibuprofen.] ▶L ♀C (D in 3rd trimester) ▶? ©II $$$

Darvocet (propoxyphene + acetaminophen): Multiple strengths; write specific product on Rx. 50/325, 2 tabs PO q4h prn. 100/500 or 100/650, 1 tab PO q4h prn. [Generic/Trade: Tabs 50/325 (Darvocet N-50), 100/650 (Darvocet N-100), & 100/500 (Darvocet A500), mg propoxyphene/acetaminophen.] ▶L ♀C ▶+ ©IV $

Darvon Compound Pulvules (propoxyphene + aspirin + caffeine, ✚*692 tablet*): 1 cap PO q4h prn. [Generic/Trade: Cap 65 mg propoxyphene/389 mg ASA/32.4 mg caffeine.] ▶LK ♀D ▶– ©IV $

Empirin with Codeine (aspirin + codeine, ✚*292 tablet*): Multiple strengths; write specific product on Rx.1-2 tabs PO q4h prn. [Generic only: Tab 325/30 & 325/60 mg ASA/mg codeine. Empirin brand no longer made.] ▶LK ♀D ▶– ©III $

Fioricet with Codeine (acetaminophen + butalbital + caffeine + codeine): 1-2 caps PO q4h prn. [Generic/Trade: Cap 325 mg acetaminophen/50 mg butalbital/40 mg caffeine/30 mg codeine.] ▶LK ♀C ▶– ©III $$$

Fiorinal with Codeine (aspirin + butalbital + caffeine + codeine, ✚*Fiorinal C-1/4, Fiorinal C-1/2, Tecnal C-1/4, Tecnal C-1/2*): 1-2 caps PO q4h prn. [Generic/Trade: Cap 325 mg ASA/50 mg butalbital /40 mg caffeine/30 mg codeine.] ▶LK ♀D ▶– ©III $$$

Lorcet (hydrocodone + acetaminophen): Multiple strengths; write specific product on Rx. 5/500: 1-2 caps PO q4-6h prn. 7.5/650 & 10/650: 1 tab PO q4-6h prn. [Generic/Trade: Caps 5/500 mg, tabs 7.5/650, 10/650 mg hydrocodone/acetaminophen.] ▶LK ♀C ▶– ©III $

Lortab (hydrocodone + acetaminophen): Multiple strengths; write specific product on Rx. 1-2 tabs PO q4-6h prn (2.5/500 & 5/500). 1 tab PO q4-6h prn (7.5/500 & 10/500). Elixir: 15 mL PO q4-6h prn. [Generic/Trade: Lortab 5/500 (scored), Lortab 7.5/500 (trade scored) & Lortab 10/500 mg hydrocodone mg acetaminophen. Elixir: 7.5/500 mg/15 mL. Trade only: Tabs Lortab 2.5/500.] ▶LK ♀C ▶– ©III $

Maxidone (hydrocodone + acetaminophen): 1 tab PO q4-6h prn, max dose 5 tabs/ day. [Trade: Tab 10/750 mg hydrocodone/acetaminophen] ▶L ♀C ▶– ©III $$$

Mersyndol with Codeine (acetaminophen + codeine + doxylamine): Canada only. 1-2 tabs PO q4-6h prn. Max 12 tabs/24 hours. [Trade only: OTC tab acetaminophen 325 mg + codeine phosphate 8 mg + doxylamine 5 mg.] ▶LK ♀C ▶? $

Norco (hydrocodone + acetaminophen): 1 tab PO q4-6h prn. [Trade: Tabs 5/325, 7.5/325 & 10/325 mg hydrocodone/acetaminophen, scored.] ▶L ♀C ▶? ©III $$$

Percocet (oxycodone + acetaminophen, ✚*Percocet-demi, Oxycocet, Endocet*): Multiple strengths; write specific product on Rx. 1-2 tabs PO q4-6h prn (2.5/325 & 5/325). 1 tab PO q4-6h prn (7.5/500 & 10/650). [Trade only: Tabs 2.5/325 oxycodone/acetaminophen. Generic/Trade: Tabs 5/325, 7.5/325, 7.5/500, 10/325, 10/650 mg. Generic only: 2.5/300, 5/300, 7.5/300, 10/300, 2.5/400, 5/400, 7.5/400, 10/400, 10/500 mg.] ▶L ♀C ▶– ©II $

Percodan (oxycodone + aspirin, ✚*Oxycodan, Endodan*): Percodan: 1 tab PO q6h prn. *Percodan Demi*: 1-2 tabs PO q6h prn. [Generic/Trade: Tab *Percodan* 4.88/

325 mg oxycodone/ASA (trade scored). Trade only: *Percodan Demi* 2.44/325 mg scored.] ▶LK ♀D ▶- ©II $$

Roxicet (oxycodone + acetaminophen): Multiple strengths; write specific product on Rx. 1 tab PO q6h prn. Soln: 5 mL PO q6h prn. [Generic/Trade: Tab 5/325 mg, scored. Cap/Caplet 5/500 mg. Trade only: Soln 5/325 per 5 mL, mg oxycodone/acetaminophen.] ▶L ♀C ▶- ©II $

Soma Compound with Codeine (carisoprodol + aspirin + codeine): Moderate to severe musculoskeletal pain:1-2 tabs PO qid prn. [Generic/Trade: Tab 200 mg carisoprodol/325 mg ASA/16 mg codeine.] ▶L ♀D ▶- ©III $$$

Synalgos-DC (dihydrocodeine + aspirin + caffeine): 2 caps PO q4h prn. [Trade only: Cap 16 mg dihydrocodeine/356.4 mg ASA/30 mg caffeine. "Painpack"=12 caps.] ▶L ♀C ▶- ©III $

Talacen (pentazocine + acetaminophen): 1 tab PO q4h prn. [Generic/Trade: Tab 25 mg pentazocine/650 mg acetaminophen, trade scored.] ▶L ♀C ▶? ©IV $$$

Tylenol with Codeine (codeine + acetaminophen): ♥*Lenoltec, Emtec, Triatec*): Multiple strengths; write specific product on Rx. 1-2 tabs PO q4h prn. Elixir 3-6 yo 5 mL/dose. 7-12 yo 10 mL/dose q4-6h prn. [Generic only: Tabs *Tylenol #2* (15/300), *Tylenol #4* (60/300). Trade only: *Tylenol with Codeine Elixir* 12/120 per 5 mL, mg codeine/mg acetaminophen. Generic/Trade: Tabs *Tylenol #3* (30/300). Canadian forms come with (*Lenoltec, Tylenol*) or without (*Empracet, Emtec*) caffeine.] ▶LK ♀C ▶? ©III (Tabs), V(elixir) $

Tylox (oxycodone + acetaminophen): 1 cap PO q6h prn. [Generic/Trade: Cap 5 mg oxycodone/500 mg acetaminophen.] ▶L ♀C ▶- ©II $

Vicodin (hydrocodone + acetaminophen): Multiple strengths; write specific product on Rx. 5/500 & 7.5/750: 1-2 tabs PO q4-6h prn. 10/660: 1 tab PO q4-6h prn. [Generic/Trade: Tabs *Vicodin* (5/500), *Vicodin ES* (7.5/750), *Vicodin HP* (10/660), scored, mg hydrocodone/mg acetaminophen.] ▶LK ♀C ▶? ©III $

Vicoprofen (hydrocodone + ibuprofen): 1 tab PO q4-6h prn. [Generic/Trade: Tab 7.5 mg hydrocodone/200 mg ibuprofen.] ▶LK ♀- ▶? ©III $$$

Wygesic (propoxyphene + acetaminophen): 1 tab PO q4h prn. [Generic/Trade: Tab 65 mg propoxyphene/650 mg acetaminophen.] ▶L ♀C ▶? ©IV $

Zydone (hydrocodone + acetaminophen): Multiple strengths; write specific product on Rx: 1-2 tabs PO q4-6h prn (5/400). 1 tab q4-6h prn (7.5/400,10/400). [Trade only: Tabs 5/400, 7.5/400, & 10/400 mg hydrocodone/acetam.] ▶LK ♀C ▶? ©III $$

Opioid Antagonists

nalmefene (*Revex*): Opioid overdose: 0.5 mg/70 kg IV. If needed, this may be followed by a second dose of 1 mg/70 kg, 2-5 minutes later. Max cumulative dose 1.5 mg/70 kg. If suspicion of opioid dependency, initially administer a challenge dose of 0.1 mg/70 kg. Post-operative opioid reversal: 0.25 mcg/kg IV followed by 0.25 mcg/kg incremental doses at 2-5 minute intervals, stopping as soon as the desired degree of opioid reversal is obtained. Max cumulative dose 1 mcg/kg. [Trade only: Injection 100 mcg/mL nalmefene for postoperative reversal (blue label). 1 mg/mL nalmefene for opioid overdose (green label).] ▶L ♀B ▶? $$$

naloxone (*Narcan*): Opioid overdose: 0.4-2.0 mg q2-3 min prn. Adult post-op reversal 0.1-0.2 mg. Ped post-op reversal: 0.005-0.01 mg. IV/IM/SC/ET. ▶LK ♀B ▶? $

Other Analgesics

acetaminophen (*Tylenol, Panadol, Tempra, paracetamol*, ♥*Abenol, Atasol, Pediatrix*): 325-650 mg PO/PR q4-6h prn. Max dose 4 g/day. OA: 2 extended release caplets (ie, 1300 mg) PO q8h around the clock. Peds: 10-15 mg/kg/dose PO/PR q4-6h prn. [OTC: Tabs 160, 325, 500, 650 mg. Chewable Tabs 80 mg. Gelcaps 500 mg. Caps 325 & 500 mg. Sprinkle Caps 80 & 160 mg. Extended-release caplets 650 mg. Liquid 160 mg/5 mL & 500 mg/15 mL. Drops 80 mg/0.8 mL. Suppositories 80, 120, 125, 300, 325, & 650 mg.] ▸LK ♀B ▸+ $

tramadol (*Ultram, Ultram ER*): Moderate to moderately severe pain: 50-100 mg PO q4-6h prn, max 400 mg/day. Adjust dose in elderly, renal & hepatic dysfunction. Avoid in opioid-dependent patients. Seizures may occur with concurrent antidepressants or seizure disorder. [Generic/Trade: Tab, immediate-release 50 mg. Trade only: Extended release tabs 100, 200, 300 mg.] ▸KL ♀C ▸- $$$

Women's Tylenol Menstrual Relief (acetaminophen + pamabrom): 2 caplets PO q4-6h. [OTC: Caplet 500 mg acet./25 mg pamabrom (diuretic).] ▸LK ♀B ▸+ $

ANESTHESIA

Anesthetics & Sedatives

dexmedetomidine (*Precedex*): ICU sedation <24h: Load 1 mcg/kg over 10 min followed by infusion 0.2-0.7 mcg/kg/h titrated to desired sedation endpoint. Beware of bradycardia and hypotension. ▸LK ♀C ▸? $$$$

etomidate (*Amidate*): Induction 0.3 mg/kg IV. ▸L ♀C ▸? $

ketamine (*Ketalar*): 1-2 mg/kg IV over 1-2 min or 4 mg/kg IM induces 10-20 min dissociative state. Concurrent atropine minimizes hypersalivation. ▸L ♀? ▸? ©III $

methohexital (*Brevital*): Induction 1-1.5 mg/kg IV, duration 5 min. ▸L ♀B ▸? ©IV $

midazolam (*Versed*): Adult sedation/anxiolysis: 5 mg or 0.07 mg/kg IM; or 1 mg IV slowly q2-3 min up to 5 mg. Peds: 0.25-1.0 mg/kg to max of 20 mg PO, or 0.1-0.15 mg/kg IM. IV route (6 mo to 5 yo): Initial dose 0.05-0.1 mg/kg IV, then titrated to max 0.6 mg/kg. IV route (6-12 yo): Initial dose 0.025-0.05 mg/kg IV, then titrated to max 0.4 mg/kg. Monitor for resp depression. [Oral liquid 2 mg/mL] ▸LK ♀D ▸- ©IV $

pentobarbital (*Nembutal*): Pediatric sedation: 1-6 mg/kg IV, adjusted in increments of 1-2 mg/kg to desired effect, or 2-6 mg/kg IM, max 100 mg. ▸LK ♀D ▸? ©II $$

propofol (*Diprivan*): 40 mg IV q10 sec until induction (2-2.5 mg/kg). ICU ventilator sedation: infusion 5-50 mcg/kg/min. ▸L ♀B ▸- $$$

thiopental (*Pentothal*): Induction 3-5 mg/kg IV, duration 5 min. ▸L ♀C ▸? ©III $

Local Anesthetics

articaine (*Septocaine, Zorcaine*): 4% injection (includes epinephrine). [4% (includes epinephrine 1:100,000)] ▸LK ♀C ▸? $

bupivacaine (*Marcaine, Sensorcaine*): Local and regional anesthesia. [0.25%, 0.5%, 0.75%, all with or without epinephrine.] ▸LK ♀C ▸? $

Duocaine (bupivacaine + lidocaine): Local anesthesia, nerve block for eye surgery. [Vials contain bupivacaine 0.375% + lidocaine 1%.] ▸LK ♀C ▸? $

lidocaine (*Xylocaine*): 0.5-1% injection with and without epinephrine. [0.5,1,1.5,2%. With epi: 0.5,1,1.5,2%.] ▸LK ♀B ▸? $

mepivacaine (*Carbocaine, Polocaine*): 1-2% injection. [1,1.5,2,3%.] ▸LK ♀C ▸? $

Neuromuscular Blockers

Should be administered only by those skilled in airway management and respiratory support.

atracurium (*Tracrium*): 0.4-0.5 mg/kg IV. Duration 15-30 min. ▸Plasma ♀C ▸? $

cisatracurium (*Nimbex*): 0.1-0.2 mg/kg IV. Duration 30-60 min. ▸Plasma ♀B ▸? $$

pancuronium (**Pavulon**): 0.04 to 0.1 mg/kg IV. Duration 45 min. ▶LK ♀C ▶? $
rocuronium (**Zemuron**): 0.6 mg/kg IV. Duration 30 min. ▶L ♀B ▶? $$
succinylcholine (**Anectine, Quelicin**): 0.6-1.1 mg/kg IV. Peds: 2 mg/kg IV, consider pretreatment with atropine 0.02 mg/kg if <5 yo. ▶Plasma ♀C ▶? $
vecuronium (**Norcuron**): 0.08-0.1 mg/kg IV. Duration 15-30 min. ▶LK ♀C ▶? $

ANTIMICROBIALS

Aminoglycosides

NOTE: See also dermatology and ophthalmology

amikacin (**Amikin**): 15 mg/kg up to 1500 mg/day IM/IV divided q8-12h. Peak 20-35 mcg/mL, trough <5 mcg/mL. Alternative 15 mg/kg IV q24h. ▶K ♀D ▶? $$$

gentamicin (**Garamycin**): Adults: 3-5 mg/kg/day IM/IV divided q8h. Peak 5-10 mcg/mL, trough <2 mcg/mL. Alternative 5-7 mg/kg IV q24h. Peds: 2-2.5 mg/kg q8h. ▶K ♀D ▶+ $$

streptomycin: Combo therapy for TB: 15 mg/kg up to 1 g IM daily. 10 mg/kg up to 750 mg if >59 yo. Peds: 20-40 mg/kg up to 1 g IM daily. Nephrotoxicity, ototoxicity. [Generic: 1 g vials for parenteral use.] ▶K ♀D ▶? $$$$$

tobramycin (**Nebcin, TOBI**): Adults: 3-5 mg/kg/day IM/IV divided q8h. Peak 5-10 mcg/mL, trough <2 mcg/mL. Alternative 5-7 mg/kg IV q24h. Peds: 2-2.5 mg/kg q8h. Cystic fibrosis (TOBI): 300 mg neb bid 28 days on, then 28 days off. [Trade only: TOBI 300 mg ampules for nebulizer.] ▶K ♀D ▶? $$$$

Antifungal Agents

amphotericin B deoxycholate (**Fungizone**): Test dose 0.1 mg/kg up to 1 mg slow IV. Wait 2-4 h, and if tolerated then begin 0.25 mg/kg IV daily and advance to 0.5-1.5 mg/kg/day depending on fungal type. Maximum dose 1.5 mg/kg/day. ▶Tissues ♀B ▶? $$$$

amphotericin B lipid formulations (**Amphotec, Abelcet, AmBisome**): Abelcet: 5 mg/kg/day IV at 2.5 mg/kg/hr. AmBisome: 3-5 mg/kg/day IV over 2 h. Amphotec: Test dose of 10 mL over 15-30 minutes, observe for 30 minutes, then 3-4 mg/kg/day IV at 1 mg/kg/h. ▶? ♀B ▶? $$$$$

anidulafungin (**Eraxis**): Candidemia: 200 mg IV load on day 1, then 100 mg IV once daily. Esophageal candidiasis: 100 mg IV load on day 1, then 50 mg IV once daily. Max infusion rate of 1.1 mg/min to prevent histamine reactions. ▶Degraded chemically ♀C ▶? $$$$$

caspofungin (**Cancidas**): 70 mg IV loading dose on day 1, then 50 mg IV daily. Infuse over 1 h. ▶KL ♀C ▶? $$$$$

clotrimazole (**Mycelex**, ✚**Canesten, Clotrimaderm**): Oral troches 5 x/day x 14 days. [Generic/Trade: Oral troches 10 mg.] ▶L ♀C ▶? $$$

fluconazole (**Diflucan**): Vaginal candidiasis: 150 mg PO single dose ($). All other dosing regimens IV/PO. Oropharyngeal/esophageal candidiasis: 200 mg first day, then 100 mg daily. Systemic candidiasis, cryptococcal meningitis: 400 mg daily. Peds: Oropharyngeal/esophageal candidiasis: 6 mg/kg first day, then 3 mg/kg daily. Systemic candidiasis: 6-12 mg/kg daily. Cryptococcal meningitis: 12 mg/kg on first day, then 6 mg/kg daily. [Generic/Trade: Tabs 50, 100, 200 mg; 150 mg tab in single-dose blister pack; susp 10 & 40 mg/mL (35 mL).] ▶K ♀C ▶+ $$$

flucytosine (**Ancobon**): 50-150 mg/kg/day PO divided qid. Myelosuppression. [Trade only: Caps 250, 500 mg.] ▶K ♀C ▶- $$$$$

griseofulvin (**Grisactin 500, Grifulvin V**, ✚**Fulvicin**): Tinea capitis: 500 mg PO daily in adults; 15-20 mg/kg up to 1 g PO daily in peds. Treat x 4-6 weeks, continuing for 2 weeks past symptom resolution. [Generic/Trade: Susp 125 mg/5 mL (120 mL), Trade only: tabs 500 mg.] ▶Skin ♀C ▶? $$$

itraconazole (Sporanox): Oral caps for onychomycosis "pulse dosing": 200 mg PO bid for 1st wk of month x 2 months (fingernails) or 3-4 months (toenails). Oral soln for oropharyngeal or esophageal candidiasis: 100-200 mg PO daily or 100 mg bid swish & swallow in 10 ml increments on empty stomach. For life-threatening infections, load with 200 mg IV bid x 4 doses or 200 mg PO tid x 3 days. Empiric therapy of suspected fungal infection in febrile neutropenia: 200 mg IV bid x 4 doses, then 200 mg IV daily for ≤14 days. Continue with oral soln 200 mg (20 ml) PO bid until significant neutropenia resolved. Contraindicated with cisapride, dofetilide, ergot alkaloids, lovastatin, PO midazolam, pimozide, quinidine, simvastatin, triazolam. Negative inotrope; do not use for onychomycosis if ventricular dysfunction. [Generic/Trade: Cap 100 mg. Trade only: Oral soln 10 mg/mL (150 mL).] ▶L ♀C ▶- $$$$$

ketoconazole (Nizoral): 200-400 mg PO daily. Hepatotoxicity. Contraindicated with cisapride, midazolam, pimozide, triazolam. H2 blockers, proton pump inhibitors, antacids impair absorption. [Generic/Trade: Tabs 200 mg.] ▶L ♀C ▶?+ $$$

micafungin (Mycamine): Infuse IV over 1 h. Esophageal candidiasis: 150 mg once daily. Prevention of candidal infections in bone marrow transplant patients: 50 mg once daily. ▶L, feces ♀C ▶? $$$$$

nystatin (Mycostatin, ♥Nilstat, Nyaderm, Candistatin): Thrush: 4-6 mL PO swish & swallow qid. Infants: 2 mL/dose with 1 mL in each cheek qid. [Generic/Trade: Susp 100,000 units/mL (60,480 mL). Trade only: Troches 200,000 units.] ▶Not absorbed ♀B ▶? $$$

terbinafine (Lamisil): Onychomycosis: 250 mg PO daily x 6 weeks for fingernails, x 12 weeks for toenails. "Pulse dosing": 500 mg PO daily for first week of month x 2 months (fingernails) or 4 months (toenails). [Trade only: Tabs 250 mg.] ▶LK ♀B ▶- $$$$$

voriconazole (Vfend): IV: 6 mg/kg q12h x 2, then 3-4 mg/kg IV q12h (use 4 mg/kg for non-candidal infections). Infuse over 2 h. PO: 200 mg q12h if >40 kg, 100 mg PO q12h if <40 kg. Take 1 h before/after meals. Treat esophageal candidiasis with oral regimen for ≥2 weeks & continuing for >1 week past symptom resolution. Treat systemic candidal infections for ≥2 weeks past symptom resolution or last positive culture, whichever is longer. Many drug interactions. [Trade only: Tabs 50,200 mg (contains lactose), susp 40 mg/mL (75mL).] ▶L ♀D ▶? $$$$$

Antimalarials

NOTE: For help treating malaria or getting antimalarials, call the CDC "malaria hotline" (770) 488-7788 Monday-Friday 8 am to 4:30 pm EST. After hours / weekend (404) 639-2888. Information is also available at: http://www.cdc.gov.

chloroquine (Aralen): Malaria prophylaxis, chloroquine-sensitive areas: 8 mg/kg up to 500 mg PO q wk from 1-2 weeks before exposure to 4 weeks after. Chloroquine resistance widespread. [Generic only: Tabs 250 mg. Generic/Trade: Tabs 500 mg (500 mg phosphate equivalent to 300 mg base).] ▶KL ♀C but + ▶+ $$

doxycycline (Adoxa, Vibramycin, Vibra-Tabs, Doryx, Monodox, ♥Doxycin): Malaria prophylaxis: 2 mg/kg/day up to 100 mg PO daily starting 1-2 days before exposure until 4 weeks after. 100 mg IV/PO bid for severe bacterial infections. Avoid in children <8 yo due to teeth staining. [Generic/Trade: Tabs 75,100 mg, caps 20, 50,100 mg. Trade only: (Vibramycin) Susp 25 mg/5 mL (60 mL), Syrup 50 mg/5 mL (480 mL). Delayed Release (Doryx): Tabs 75,100 mg. Generic: Caps 75,150 mg tabs 50,100 mg.] ▶LK ♀D ▶? $

Fansidar (sulfadoxine + pyrimethamine): Chloroquine-resistant P falciparum malaria: 2-3 tabs PO single dose. Peds: 5-10 kg: ½ tab. 11-20 kg: 1 tab. 21-30 kg: 1½ tab. 31-45 kg: 2 tabs. >45 kg: 3 tabs. Do not use in infants <2 mo; may cause kernicterus. Stevens-Johnson syndrome, toxic epidermal necrolysis. *Fansidar* resistance common in many malarious areas. [Trade only: Tabs sulfadoxine 500 mg + pyrimethamine 25 mg.] ▶KL ♀C ▶- $

Malarone (atovaquone + proguanil): Prevention of malaria: 1 adult tab PO daily from 1-2 days before exposure until 7 days after. Treatment of malaria: 4 adult tabs PO daily x 3 days. Take with food or milky drink. [Trade only: Adult tabs atovaquone 250 mg + proguanil 100 mg; pediatric tabs 62.5 mg + 25 mg.] ▶Fecal excretion; LK ♀C ▶? $$$$

mefloquine (*Lariam*): Malaria prophylaxis for chloroquine-resistant areas: 250 mg PO q week from 1 week before exposure to 4 weeks after. Treatment: 1250 mg PO single dose. Peds: Malaria prophylaxis: Give PO once weekly starting 1 week before exposure to 4 weeks after: <15 kg, 5 mg/kg (prepared by pharmacist); 15-19 kg, ¼ tab; 20-30 kg, ½ tab; 31-45 kg, ¾ tab; >45 kg, 1 tab. Treatment: 20-25 mg/kg PO; can divide into 2 doses given 6-8 h apart. Take on full stomach. [Generic/Trade: Tabs 250 mg.] ▶L ♀C ▶? $$

primaquine: 30 mg base PO daily x 14 days. [Generic only: Tabs 26.3 mg (equiv to 15 mg base).] ▶L ♀- ▶- $

quinidine: Life-threatening malaria: Load with 10 mg/kg (max 600 mg) IV over 1-2 h, then 0.02 mg/kg/min. Treat x 72 h, until parasitemia <1%, or PO meds tolerated. Dose given as quinidine gluconate. ▶LK ♀C ▶? $$$$

quinine: Malaria: 600-650 mg PO tid. Peds: 25-30 mg/kg/day up to 2 g/day PO divided q8h. Treat for 3-7 days. Also give doxycycline or Fansidar. [Generic: Tabs 260 mg, caps 200,325 mg.] ▶L ♀X ▶+? $

Antimycobacterial Agents

> **NOTE:** Two or more drugs are needed for the treatment of active mycobacterial infections. See guidelines at http://www.thoracic.org/statements/.

dapsone: Pneumocystis prophylaxis, leprosy: 100 mg PO daily. Pneumocystis treatment: 100 mg PO daily with trimethoprim 5 mg/kg PO tid x 21 days. [Generic: Tabs 25,100 mg.] ▶LK ♀C ▶+? $

ethambutol (*Myambutol*, ✚*Etibi*): 15-20 mg/kg PO daily. Dose with whole tabs: Give PO daily 800 mg if 40-55 kg, 1200 mg if 56-75 kg, 1600 mg if 76-90 kg. Base dose on estimated lean body weight. Peds: 15-20 mg/kg up to 1 g PO daily. [Generic/Trade: Tabs 100,400 mg.] ▶LK ♀C but + ▶+ $$$$

isoniazid (*INH*, ✚*Isotamine*): Adults: 5 mg/kg up to 300 mg PO daily. Peds: 10-15 mg/kg up to 300 mg PO daily. Hepatotoxicity. Consider supplemental pyridoxine 10-50 mg PO daily. [Generic: Tabs 100,300 mg, syrup 50 mg/5 mL.] ▶LK ♀C but + ▶+ $

pyrazinamide (*PZA*, ✚*Tebrazid*): 20-25 mg/kg up to 2000 mg PO daily. Dose with whole tabs: Give PO daily 1000 mg if 40-55 kg, 1500 mg if 56-75 kg, 2000 mg if 76-90 kg. Base dose on estimated lean body weight. Peds: 15-30 mg/kg up to 2000 mg PO daily. Hepatotoxicity. [Generic: Tabs 500 mg.] ▶LK ♀C ▶? $$$

rifabutin (*Mycobutin*): 300 mg PO daily or 150 mg PO bid. [Trade only: Caps 150 mg.] ▶L ♀B ▶? $$$$$

Rifamate (isoniazid + rifampin): 2 caps PO daily. [Trade only: Caps isoniazid 150 mg + rifampin 300 mg.] ▶LK ♀C but + ▶+ $$$$

rifampin (*Rimactane*, *Rifadin*, ✚*Rofact*): TB: 10 mg/kg up to 600 mg PO/IV daily.

Peds: 10-20 mg/kg up to 600 mg PO/IV daily. IV and PO doses are the same. [Generic/Trade: Caps 150,300 mg. Pharmacists can make oral suspension.] ▶L ♀C but + ▶+ $$$$

rifapentine (**_Priftin_**): 600 mg PO twice weekly x 2 months, then once weekly x 4 months. Use for continuation therapy only in selected HIV-negative patients. [Trade only: Tabs 150 mg.] ▶Esterases, fecal ♀C ▶? $$$$

Rifater (isoniazid + rifampin + pyrazinamide): 6 tabs PO daily if ≥55 kg, 5 daily if 45-54 kg, 4 daily if ≤44 kg. [Trade only: Tab Isoniazid 50 mg + rifampin 120 mg + pyrazinamide 300 mg.] ▶LK ♀C ▶? $$$$$

Antiparasitics

albendazole (**_Albenza_**): Hydatid disease, neurocysticercosis: 400 mg PO bid. 15 mg/kg/day up to 800 mg/day if <60 kg. [Trade only: Tabs 200 mg.] ▶L ♀C ▶? $$$

atovaquone (**_Mepron_**): Pneumocystis treatment: 750 mg PO bid x 21 days. Pneumocystis prevention: 1500 mg PO daily. Take with meals. [Trade only: Susp 750 mg/5 mL, foil pouch 750 mg/5 mL.] ▶Fecal ♀C ▶? $$$$$

ivermectin (**_Stromectol_**): Single PO dose of 200 mcg/kg for strongyloidiasis, scabies (not for children <15 kg), 150 mcg/kg for onchocerciasis. Take on empty stomach with water. [Trade only: Tab 3, 6 mg.] ▶L ♀C ▶+ $

mebendazole (**_Vermox_**): Pinworm: 100 mg PO x 1; repeat in 2 wks. Roundworm, whipworm, hookworm: 100 mg PO bid x 3d. [Generic/Trade: Chew tab 100 mg.] ▶L♀C ▶? $$

metronidazole (**_Flagyl_**, ♣**_Trikacide, Florazole ER, Nidazol_**): Trichomoniasis: 2g PO single dose for patient & sex partners. Giardia: 250 mg (5 mg/kg/dose for peds) PO tid x 5-7 days. [Generic/Trade: Tabs 250,500 mg, ER tabs 750 mg, Caps 375 mg.] ▶KL ♀B ▶?- $

nitazoxanide (**_Alinia_**): Cryptosporidial or Giardial diarrhea: 100 mg bid for 1-3 yo, 200 mg bid for 4-11 yo, 500 mg bid for adults and children ≥12 yo. Give PO with food x 3 days. Use susp for <12 yo. [Trade only: Oral susp 100 mg/5 mL 60 mL bottle, tab 500 mg.] ▶L ♀B ▶? $$$

paromomycin (**_Humatin_**): 25-35 mg/kg/day PO divided tid with or after meals. [Generic/Trade: Caps 250 mg.] ▶Not absorbed ♀C ▶- $$$

pentamidine (**_Pentam, NebuPent_**, ♣**_Pentacarinat_**): Pneumocystis treatment: 4 mg/kg IM/IV daily x 21 days. Pneumocystis prevention: 300 mg nebulized q 4 weeks. [Trade only: Aerosol 300 mg.] ▶K ♀C ▶- $$$

praziquantel (**_Biltricide_**): Schistosomiasis: 20 mg/kg PO q4-6h x 3 doses. Neurocysticercosis: 50 mg/kg PO divided tid x 15 days (up to 100 mg/kg/day for peds). [Trade only: Tabs 600 mg.] ▶LK ♀B ▶- $$$

pyrantel (**_Antiminth, Pin-X, Pinworm_**, ♣**_Combantrin_**): Pinworm and roundworm: 11 mg/kg up to 1 g PO single dose. [OTC: Caps 62.5 mg, liquid 50 mg/mL.] ▶Not absorbed ♀- ▶? $

pyrimethamine (**_Daraprim_**): CNS toxoplasmosis in AIDS. Acute therapy: 200 mg PO x 1, then 50 mg (<60 kg) to 75 mg (≥60 kg) PO once daily + sulfadiazine + leucovorin 10-20 mg PO once daily (can increase to ≥50 mg/day) for ≥6 weeks. Secondary prevention: Pyrimethamine 25-50 mg PO once daily + sulfadiazine + leucovorin 10-25 mg PO once daily. [Trade only: Tabs 25 mg.] ▶L ♀C ▶+ $$

thiabendazole (**_Mintezol_**): Helminths: 22 mg/kg/dose up to 1500 mg PO bid after meals. Treat x 2 days for strongyloidiasis, cutaneous larva migrans. [Trade only: Chew tab 500 mg, susp 500 mg/5 mL (120 mL).] ▶LK ♀C ▶? $

tinidazole (**_Tindamax_**): Adults: 2 g PO daily x 1 day for trichomoniasis or giardiasis, x 3 days for amebiasis. Peds, >3 yo: 50 mg/kg (up to 2 g) PO daily x 1 day for giardiasis, x 3 days for amebiasis. Take with food. [Trade only: Tabs 250,500 mg. Pharmacists can compound oral suspension.] ▶KL ♀C ▶?- $

Antiviral Agents - Anti-CMV

cidofovir (**Vistide**): CMV retinitis in AIDS: 5 mg/kg IV q wk x 2, then 5 mg/kg q2 wks. Severe nephrotoxicity. ▶K ♀C ▶- $$$$$

foscarnet (**Foscavir**): CMV retinitis: 60 mg/kg IV (over 1 h) q8h or 90 mg/kg IV (over 1.5-2 h) q12h x 2-3 weeks, then 90-120 mg/kg/day IV over 2h. HSV infection: 40 mg/kg (over 1 h) q8-12h. Nephrotoxicity, seizures. ▶K ♀C ▶? $$$$$

ganciclovir (**DHPG, Cytovene**): CMV retinitis: Induction 5 mg/kg IV q12h for 14-21 days. Maintenance 6 mg/kg IV daily for 5 days per week. Myelosuppression. Potential carcinogen, teratogen. May impair fertility. [Generic/Trade: Caps 250, 500 mg.] ▶K ♀C ▶- $$$$$

valganciclovir (**Valcyte**): CMV retinitis: 900 mg PO bid x 21 days, then 900 mg PO daily. Prevention of CMV disease in high-risk kidney, kidney-pancreas, heart transplant patients: 900 mg PO daily from within 10 days after transplant until 100 days post-transplant. Greater bioavailability than oral ganciclovir. Give with food. Impaired fertility, myelosuppression, potential carcinogen & teratogen. [Trade only: Tabs 450 mg.] ▶K ♀C ▶- $$$$$

Antiviral Agents - Anti-Herpetic

acyclovir (**Zovirax**): Genital herpes: 400 mg PO tid x 7-10 days for first episode, x 5 days for recurrent episodes. Chronic suppression of genital herpes: 400 mg PO bid, 400-800 mg PO bid-tid in HIV infection. Zoster: 800 mg PO 5 times/day x 7-10 days. Chickenpox: 20 mg/kg up to 800 mg PO qid x 5 days. Adult IV: 5-10 mg/kg IV q8h, each dose over 1h. Peds, herpes encephalitis: 20 mg/kg IV q8h x 10 days for 3 mo-12 yo, adult dose for ≥12 yo. [Generic/Trade: Caps 200 mg, tabs 400,800 mg. Susp 200 mg/5 mL.] ▶K ♀B ▶+ $

famciclovir (**Famvir**): First-episode genital herpes: 250 mg PO tid x 7-10 days. Recurrent genital herpes: 1000 mg PO bid x 2 doses; 500 bid x 7 days if HIV-infected. Chronic suppression of genital herpes: 250 mg PO bid; 500 mg PO bid if HIV infected. Recurrent herpes labialis: 1500 mg PO single dose; 500 bid x 7 days if HIV-infected. Zoster: 500 mg PO tid for 7 days. [Trade only: Tabs 125,250, 500 mg. All tabs contain lactose.] ▶K ♀B ▶? $$

valacyclovir (**Valtrex**): First-episode genital herpes: 1000 mg PO bid x 10 days. Recurrent genital herpes: 500 mg PO bid x 3 days, 1 g PO bid x 5-10 days in HIV infection. Chronic suppression of genital herpes: 500-1000 mg PO daily; 500 mg PO bid if HIV infection. Reduction of genital herpes transmission in immunocompetent patients with ≤9 recurrences/year: 500 mg PO daily by source partner, in conjunction with safer sex practices. Herpes labialis: 2 g PO q12h x 2 doses. Zoster: 1000 mg PO tid x 7 days. [Trade only: Tabs 500,1000 mg.] ▶K ♀B ▶+ $$$$

NOTE FOR ALL ANTI-HIV DRUGS: Many serious drug interactions; always check before prescribing! AIDS treatment guidelines available online at www.aidsinfo.nih.gov.

Antiviral Agents - Anti-HIV - Combinations

Atripla (efavirenz + emtricitabine + tenofovir): 1 tab PO once daily on empty stomach, preferably at bedtime. [Trade only: Tabs efavirenz 600 mg + emtricitabine 200 mg + tenofovir 300 mg.] ▶KL ♀+ ▶- $$$$$

Epzicom (abacavir + lamivudine): 1 tab PO daily. [Trade only: Tabs abacavir 600 mg + lamivudine 300 mg.] ▶LK ♀C ▶- $$$$$

Trizivir (abacavir + lamivudine + zidovudine): 1 tab PO bid. [Trade only: Tabs abacavir 300 mg + lamivudine 150 mg + zidovudine 300 mg.] ▶LK ♀C ▶- $$$$$

Truvada (emtricitabine + tenofovir): 1 tab PO daily. [Trade only: Tabs emtricitabine 200 mg + tenofovir 300 mg.] ▶K ♀B ▶- $$$$$

Antiviral Agents - Anti-HIV - Fusion Inhibitors

enfuvirtide (*Fuzeon, T-20*): 90 mg SC bid. Peds, ≥6 yo: 2 mg/kg up to 90 mg SC bid. [30-day kit with vials, diluent, syringes, alcohol wipes. Single-dose vials contain 108 mg to provide 90 mg enfuvirtide.] ▶Serum ♀B ▶- $$$$$

Antiviral Agents - Anti-HIV - Non-Nucleoside Reverse Transcript. Inhibitors

efavirenz (*Sustiva, EFV*): Adults & children >40 kg: 600 mg PO qhs. Peds, >=3 yo: Give PO qhs 200 mg for 10-15 kg; 250 mg for 15-20 kg; 300 mg for 20 to <25 kg; 350 mg for 25 to <32.5 kg; 400 mg for 32.5 to <40 kg. Do not give with high-fat meal. [Trade only: Caps 50,100,200 mg, tabs 600 mg.] ▶L ♀D ▶- $$$$$

nevirapine (*Viramune, NVP*): 200 mg PO daily x 14 days initially. If tolerated, increase to 200 mg PO bid. Peds <8 yo: 4 mg/kg PO daily x 14 days, then 7 mg/kg PO bid. ≥8 yo: 4 mg/kg PO daily x 14 days, then 4 mg/kg PO bid. Severe skin reactions & hepatotoxicity. [Trade only: Tabs 200 mg, susp 50 mg/5 mL (240 mL).] ▶LK ♀C ▶- $$$$$

Antiviral Agents - Anti-HIV - Nucleoside/Nucleotide Reverse Transcrip Inhib

abacavir (*Ziagen, ABC*): Adult: 300 mg PO bid or 600 mg PO daily. Children >3 mo: 8 mg/kg up to 300 mg PO bid. Potentially fatal hypersensitivity; never rechallenge if this occurs. Severe hypersensitivity may be more common with once-daily regimen. [Trade: Tabs 300 mg, oral soln 20 mg/mL (240 mL).] ▶L ♀C ▶- $$$$$

Combivir (lamivudine + zidovudine): 1 tab PO bid. [Trade only: Tabs lamivudine 150 mg + zidovudine 300 mg.] ▶LK ♀C ▶- $$$$$

didanosine (*Videx, Videx EC, ddI*): Videx EC, adults: 400 mg PO daily if ≥60 kg, 250 mg PO daily if <60 kg. Dosage reduction of Videx EC with tenofovir: 250 mg if ≥60 kg, 200 mg if <60 kg. Dosage reduction unclear with tenofovir if CrCl <60 mL/min. Buffered powder, peds: 100 mg/m² PO bid for age 2 wks-8 mo. 120 mg/m² PO bid for >8 mo. All forms usually taken on empty stomach. [Trade only: Packets of buffered powder for oral soln 100,167,250 mg. Peds powder for oral soln 10 mg/mL (buffered with antacid). Generic/Trade: Delayed-release caps 200,250, 400 mg. Trade only: (Videx EC) delayed-release caps 125 mg.] ▶LK ♀B ▶- $$$$$

emtricitabine (*Emtriva, FTC*): 200 mg cap or 240 mg oral soln PO once daily. Peds ≥3 mo: 6 mg/kg (max 240 mg) oral soln PO once daily. Can give 200 mg cap PO once daily if >33 kg. [Trade only: Caps 200 mg, oral soln 10 mg/mL (170 mL).] ▶K ♀B ▶- $$$$$

lamivudine (*Epivir, Epivir-HBV, 3TC, ♣Heptovir*): Epivir for HIV infection. Adults: 150 mg PO bid or 300 mg PO daily. Peds: 4 mg/kg up to 150 mg PO bid. Epivir-HBV for hepatitis B: Adults: 100 mg PO daily. Peds: 3 mg/kg up to 100 mg PO daily. [Trade only: 3TC: Tabs 150, 300 mg, oral soln 10 mg/mL. Epivir-HBV, Heptovir: Tabs 100 mg, oral soln 5 mg/mL.] ▶K ♀C ▶- $$$$$

stavudine (*Zerit, d4T*): 40 mg PO q12h, or 30 mg q12h if <60 kg. Peds (<30 kg): 1 mg/kg PO q12h. [Trade only: Caps 15,20,30,40 mg; oral soln 1 mg/mL (200 mL).] ▶LK ♀C ▶- $$$$$

tenofovir (**Viread, TDF**): 300 mg PO daily with a meal. [Trade only: Tab 300 mg.] ▶K ♀B ▶- $$$$$

zidovudine (**Retrovir, AZT, ZDV**): 200 mg PO tid or 300 bid. Peds: 160 mg/m^2 up to 200 mg PO q8h. [Generic/Trade: Tab 300 mg, syrup 50 mg/5 mL (240 mL). Trade: Cap 100 mg.] ▶LK ♀C ▶- $$$$$

Antiviral Agents - Anti-HIV - Protease Inhibitors

atazanavir (**Reyataz, ATV**): Therapy-naïve patients: 400 mg PO daily. Efavirenz regimen, therapy-naïve patients: atazanavir 300 mg + ritonavir 100 mg + efavirenz 600 mg all PO daily. Therapy-experienced patients: atazanavir 300 mg PO daily + ritonavir 100 mg PO daily. Tenofovir regimen (must include ritonavir): atazanavir 300 mg + ritonavir 100 mg + tenofovir 300 mg all PO daily with food. Give atazanavir with food; give 2 h before or 1 h after buffered didanosine. [Trade only: Caps 100,150,200 mg.] ▶L ♀B ▶- $$$$$

darunavir (**Prezista**): Therapy-experienced patients: 600 mg boosted by ritonavir 100 mg PO bid with food. Do not use without ritonavir. [Trade only: Tab 300 mg.] ▶L ♀B ▶- $$$$$

fosamprenavir (**Lexiva, 908**): Therapy-naïve patients: 1400 mg PO bid (without ritonavir). OR fosamprenavir 1400 mg + ritonavir 200 mg both PO bid. OR 700 mg fosamprenavir + 100 mg ritonavir both PO bid. Protease inhibitor-experienced patients: 700 mg fosamprenavir + 100 mg ritonavir both PO bid. Do not use once-daily regimen. If once-daily ritonavir-boosted regimen given with efavirenz, increase ritonavir to 300 mg/day; no increase of ritonavir dose needed for bid regimen with efavirenz. No meal restrictions. [Trade only: Tabs 700 mg (equivalent to amprenavir 600 mg).] ▶L ♀C ▶- $$$$$

indinavir (**Crixivan, IDV**): 800 mg PO q8h between meals with water (at least 48 ounces/day to prevent kidney stones). [Trade only: Caps 100,200,333,400 mg.] ▶LK ♀C ▶- $$$$$

lopinavir-ritonavir (**Kaletra, LPV/r**): Adults: Give tabs without regard to meals; give oral soln with food. Therapy-naïve patients: 2 tabs PO bid or 4 tabs once daily. 5 mL PO bid or 10 mL once daily of oral soln. Increase oral soln to 6.5 ml PO bid (not once daily) with efavirenz, nevirapine, amprenavir, or nelfinavir (dosage increase not needed with tabs). Therapy-experienced patients: No once-daily regimen. 2 tabs or 5 mL oral soln bid. Increase oral soln to 6.5 ml PO bid with efavirenz, nevirapine, amprenavir, or nelfinavir. Consider 3 tabs bid with efavirenz, nevirapine, fosamprenavir without ritonavir, or nelfinavir if reduced lopinavir susceptibility suspected. Peds: Oral soln: 6 mo-12 yo: lopinavir 12 mg/kg PO bid for 7 to <15 kg, 10 mg/kg PO bid for 15-40 kg, adult dose for >40 kg. With efavirenz, nevirapine, or amprenavir increase to 13 mg/kg PO bid for 7 to <15 kg, 11 mg/kg PO bid for 15-45 kg, adult dose for >45 kg. [Trade only: Tabs 200/50 mg lopinavir-ritonavir. Oral soln 80/20 mg/ml (160 mL).] ▶L ♀C ▶- $$$$$

nelfinavir (**Viracept, NFV**): 750 mg PO tid or 1250 mg PO bid. Peds: 20-45 mg/kg PO tid. Take with meals. [Trade only: Tab 250, 625 mg, powder 50 mg/g (114 g).] ▶L ♀B ▶- $$$$$

ritonavir (**Norvir, RTV**): Full-dose regimen (600 mg PO bid) poorly tolerated. Adult doses of 100 mg PO bid to 400 mg PO bid used to boost levels of other protease inhibitors. Peds >1 month old: Start with 250 mg/m^2 and increase q2-3 days by 50 mg/m^2 twice daily. Usual dose is 350-450 mg/m^2 to max of 600 mg PO bid. If 400 mg/m^2 twice daily not tolerated, consider other alternatives. [Trade only: Cap 100 mg, oral soln 80 mg/mL (240 mL).] ▶L ♀B ▶- $$$$$

saquinavir (*Invirase*, **SQV**): Take with/after meals. Regimens must contain ritonavir. *Invirase* 1000 mg + ritonavir 100 mg both PO bid within 2 h after meals. *Invirase* 1000 mg PO + *Kaletra* 400/100 mg both PO bid. [Trade only: *Invirase* (hard gel) caps 200 mg, tabs 500 mg.] ▶L ♀B ▶? $$$$$

tipranavir (*Aptivus*): 500 mg boosted by ritonavir 200 mg PO bid with food. Hepatotoxicity. [Trade only: Caps 250 mg.] ▶Feces ♀C ▶- $$$$$

Antiviral Agents - Anti-Influenza

amantadine (*Symmetrel*, **✦***Endantadine*): Influenza A: 100 mg PO bid. Elderly: 100 mg daily. Peds: 5 mg/kg up to 150 mg/day. [Generic: Cap 100 mg. Generic/ Trade: Tab 100 mg, syrup 50 mg/5 mL (120 mL).] ▶K ♀C ▶? $

oseltamivir (*Tamiflu*): 75 mg PO bid x 5 days starting within 2 days of symptom onset. 75 mg PO daily for prophylaxis. Peds ≥1 yo: Each dose is 30 mg if ≤15 kg, 45 mg if 16-23 kg, 60 mg if 24-40 kg, 75 mg if >40 kg or ≥13 yo. For treatment, give twice daily x 5 days starting within 2 days of sx onset. For prophylaxis, give once daily x 10 days starting within 2 days of exposure. Take with food to improve tolerability. [Trade only: Caps 75 mg, susp 12 mg/mL (25 mL).] ▶LK ♀C ▶? $$$

rimantadine (*Flumadine*): Influenza A: 100 mg PO bid x 7 days. Peds: 5 mg/kg PO daily up to 150 mg/day. [Trade only: Tabs 100 mg, syrup 50 mg/5 mL (240 mL).] ▶LK ♀C ▶- $$

zanamivir (*Relenza*): Influenza treatment, ≥7 yo: 2 puffs bid x 5 days. Influenza prevention, ≥5 yo: 2 puffs once daily x 10 days. [Trade only: Rotadisk inhaler 5 mg/puff (20 puffs).] ▶K ♀C ▶? $$

Antiviral Agents - Other

adefovir (*Hepsera*): Chronic hepatitis B: 10 mg PO daily. Nephrotoxic; lactic acidosis and hepatic steatosis; discontinuation may exacerbate hepatitis B; HIV resistance in untreated HIV infection. [Trade only: Tabs 10 mg.] ▶K ♀C ▶- $$$$$

entecavir (*Baraclude*): Chronic hepatitis B: 0.5 mg PO once daily if treatment naïve; 1 mg if lamivudine-resistant or history of viremia despite lamivudine treatment. Give 2h after last meal and 2h before next meal. [Trade only: 0.5, 1 mg tabs, solution 0.05 mg/mL (210 mL).] ▶K ♀C ▶- $$$$$

interferon alfa-2b (*Intron A*): Chronic hepatitis B: 5 million units/day or 10 million units 3 times/week SC/IM x 4 mo. Chronic hepatitis C: 3 million units SC/IM 3 times/week x 4 mo. Continue for 18-24 mo if ALT normalized. [Trade only: Powder/soln for injection 3,5,10 million units/vial. Soln for injection 18,25 million units/ multidose vial. Multidose injection pens 3,5,10 million units/dose (6 doses/pen).] ▶K? ♀C ▶?+ $$$$$

interferon alfacon-1 (*Infergen*): Chronic hepatitis C: 9 mcg SC 3 times/week x 24 weeks. If relapse/no response, increase to 15 mcg SC 3 times/week. If intolerable adverse effects, reduce to 7.5 mcg SC 3 times/week. [Trade only: Vials injectable soln 9,15 mcg.] ▶Plasma ♀C ▶? $$$$$

palivizumab (*Synagis*): Prevention of respiratory syncytial virus pulmonary disease in high-risk children: 15 mg/kg IM q month during RSV season. ▶L ♀C ▶? $$$$$

peginterferon alfa-2a (*Pegasys*): Chronic hepatitis C: 180 mcg SC in abdomen or thigh once weekly for 48 weeks +/- PO ribavirin. Hepatitis B: 180 mcg SC in abdomen or thigh once weekly for 48 weeks. May cause or worsen severe autoimmune, neuropsychiatric, ischemic, & infectious diseases. Frequent clinical & lab monitoring. [Trade only: 180 mcg/1 mL solution in single-use vial, 180 mcg/0.5 mL prefilled syringe.] ▶LK ♀C ▶- $$$$$

peginterferon alfa-2b (**PEG-Intron**): Chronic hepatitis C: Give SC once weekly for 1 year. Monotherapy 1 mcg/kg/week. In combo with oral ribavirin: 1.5 mcg/kg/week. May cause or worsen severe autoimmune, neuropsychiatric, ischemic, & infectious diseases. Frequent clinical & lab monitoring. [Trade only: 50,80,120,150 mcg/0.5 mL single-use vials with diluent, 2 syringes, and alcohol swabs. Disposable single-dose Redipen 50,80,120,150 mcg.] ▶K? ♀C ▶- $$$$$

Rebetron (interferon alfa-2b + ribavirin): Chronic hepatitis C: Interferon alfa-2b 3 million units SC 3 times/week and ribavirin (Rebetol) 600 mg PO bid if >75 kg; 400 mg q am and 600 q pm if ≤75 kg. Ribavirin dose adjusted according to hemoglobin level. Contraindicated in pregnant women or their male partners. [Trade only: Each kit contains a 2-week supply of interferon alfa-2b, ribavirin caps 200 mg. Rebetol oral soln also 40 mg/mL available.] ▶K ♀X ▶- $$$$$

ribavirin - inhaled (**Virazole**): Severe respiratory syncytial virus infection in children: Aerosol 12-18 h/day x 3-7 days. Beware of sudden pulmonary deterioration; ventilator dysfunction due to drug precipitation. ▶Lung ♀X ▶- $$$$$

ribavirin - oral (**Rebetol, Copegus, Ribasphere**): Hepatitis C. Rebetol: In combo with interferon alfa 2b (Intron A): 600 mg PO bid if >75 kg; 400 mg q am and 600 q pm if ≤75 kg. In combo with peginterferon alfa 2b (PEG-Intron): 400 mg PO bid. Copegus: In combo with peginterferon alfa 2a (Pegasys): For genotype 1/4, 1200 mg/day if ≥75 kg; 1000 mg/day if <75 kg. For genotype 2/3, 800 mg/day PO. For patients coinfected with HIV, Copegus dose is 800 mg/day regardless of genotype. Give bid with food. Decrease ribavirin dose if Hb decreases. [Generic/Trade: Caps 200 mg, tabs 200,400 mg. Generic tabs 600mg. Trade only (Rebetol): oral soln 40mg/mL (100ml).] ▶Cellular, K ♀X ▶- $$$$$

Carbapenems

ertapenem (**Invanz**): 1 g IV/IM q24h. Peds: 15 mg/kg IV/IM q12h (max 1 g/day). Infuse IV over 30 minutes. ▶K ♀B ▶? $$$$$

imipenem-cilastatin (**Primaxin**): 250-1000 mg IV q6-8h. Peds >3 mo: 15-25 mg/kg IV q6h. ▶K ♀C ▶? $$$$$

meropenem (**Merrem IV**): Complicated skin infections 10 mg/kg up to 500 mg IV q8h. Intra-abdominal infections: 20 mg/kg up to 1 g IV q8h. Peds meningitis: 40 mg/kg IV q8h for age ≥3 mo; 2 g IV q8h if >50 kg. ▶K ♀B ▶? $$$$$

Cephalosporins - 1st Generation

cefadroxil (**Duricef**): 1-2 g/day PO divided daily-bid. Peds: 30 mg/kg/day divided bid. [Generic/Trade: Tabs 1 g, caps 500 mg, susp 125,250, & 500 mg/5 mL.] ▶K ♀B ▶+ $$

cefazolin (**Ancef, Kefzol**): 0.5-1.5 g IM/IV q6-8h. Peds: 25-50 mg/kg/day divided q6-8h, severe infections 100 mg/kg/day. ▶K ♀B ▶? $$$

cephalexin (**Keflex, Panixine DisperDose**): 250-500 mg PO qid. Peds 25-50 mg/kg/day. Not for otitis media, sinusitis. [Generic/Trade: Caps 250, 500 mg. Generic only: Tabs 250, 500 mg, susp 100,125 & 250 mg/5 mL. Panixine DisperDose 125, 250 mg scored tabs for oral susp. Trade only: Caps 333, 750 mg.] ▶K ♀B ▶? $$$

Cephalosporins - 2nd Generation

cefaclor (**Ceclor, Raniclor**): 250-500 mg PO tid. Peds: 20-40 mg/kg/day PO divided tid. Otitis media: 40 mg/kg/day PO divided bid. Group A streptococcal pharyngitis: 20 mg/kg/day PO divided bid. Extended release (Ceclor CD): 375-500 mg PO bid. Serum sickness-like reactions with repeated use. [Generic/Trade: Caps 250,500

OVERVIEW OF BACTERIAL PATHOGENS (selected)

Gram Positive Aerobic Cocci: *Staph epidermidis* (coagulase negative), *Staph aureus* (coagulase positive), Streptococci: *S pneumoniae* (pneumococcus), *S pyogenes* (Group A), *S agalactiae* (Group B), enterococcus

Gram Positive Aerobic / Facultatively Anaerobic Bacilli: *Bacillus, Corynebacterium diphtheriae, Erysipelothrix rhusiopathiae, Listeria monocytogenes, Nocardia*

Gram Negative Aerobic Diplococci: *Moraxella catarrhalis, Neisseria gonorrhoeae, Neisseria meningitidis*

Gram Negative Aerobic Coccobacilli: *Haemophilus ducreyi, Haemoph. influenzae*

Gram Negative Aerobic Bacilli: *Acinetobacter, Bartonella species, Bordetella pertussis, Brucella, Burkholderia cepacia, Campylobacter, Francisella tularensis, Helicobacter pylori, Legionella pneumophila, Pseudomonas aeruginosa, Stenotrophomonas maltophilia, Vibrio cholerae, Yersinia*

Gram Neg Facultatively Anaerobic Bacilli: *Aeromonas hydrophila, Eikenella corrodens, Pasteurella multocida,* Enterobacteriaceae: *E coli, Citrobacter, Shigella, Salmonella, Klebsiella, Enterobacter, Hafnia, Serratia, Proteus, Providencia*

Anaerobes: *Actinomyces, Bacteroides fragilis, Clostridium botulinum, Clostridium difficile, Clostridium perfringens, Clostridium tetani, Fusobacterium, Lactobacillus, Peptostreptococcus*

Defective Cell Wall Bacteria: *Chlamydia pneumoniae, Chlamydia psittaci, Chlamydia trachomatis, Coxiella burnetii, Myoplasma pneumoniae, Rickettsia prowazekii, Rickettsia rickettsii, Rickettsia typhi, Ureaplasma urealyticum*

Spirochetes: *Borrelia burgdorferi, Leptospira, Treponema pallidum*

Mycobacteria: *M avium complex, M kansasii, M leprae, M tuberculosis*

mg, susp 125,187,250, 375 mg/5 mL. Extended release 500 mg. Trade only (Ceclor CD): 375 mg. Generic: Chew tabs 125,187,250,375 mg.] ▶K ♀B ▶? $$$
cefoxitin (**Mefoxin**): 1-2 g IM/IV q6-8h. Peds: 80-160 mg/kg/day IV divided q4-8h. ▶K ♀B ▶+ $$$$$
cefprozil (**Cefzil**): 250-500 mg PO bid. Peds otitis media: 15 mg/kg/dose PO bid. Peds group A streptococcal pharyngitis (second-line to penicillin): 7.5 mg/kg/dose PO bid x 10d. [Generic/Trade: Tabs 250,500 mg. Susp 125 & 250 mg/5 mL.] ▶K ♀B ▶+ $$$$
cefuroxime (**Zinacef, Ceftin, Kefurox**): 750-1500 mg IM/IV q8h. Peds: 50-100 mg/kg/day IV divided q6-8h, not for meningitis. 250-500 mg PO bid. Peds: 20-30 mg/kg/day susp PO divided bid. [Generic/Trade: Tabs 250,500 mg. Generic: 125 mg tab. Trade only: Susp 125 & 250 mg/5 mL.] ▶K ♀B ▶? $$$
loracarbef (**Lorabid**): 200-400 mg PO bid. Peds: 30 mg/kg/day for otitis media (15 mg/kg/day for other infections) divided bid. [Trade only: Caps 200,400 mg, susp 100 & 200 mg/5 mL.] ▶K ♀B ▶? $$$$

Cephalosporins - 3rd Generation

cefdinir (**Omnicef**): 14 mg/kg/day up to 600 mg/day PO divided daily or bid. [Generic/Trade: Cap 300 mg, susp 125 mg/5 mL. Trade only: susp 250 mg/5 mL.] ▶K ♀B ▶? $$$$
cefditoren (**Spectracef**): 200-400 mg PO bid with food. [Trade only: Tabs 200 mg.] ▶K ♀B ▶? $$$
cefixime (**Suprax**): 400 mg PO once daily. Gonorrhea: 400 mg PO single-dose. Peds: 8 mg/kg/day divided daily-bid. [Trade only: Tabs 400 mg, susp 100 mg/5 mL.] ▶K/Bile ♀B ▶? $

SEXUALLY TRANSMITTED DISEASES & VAGINITIS*

Bacterial vaginosis: 1) metronidazole 5 g of 0.75% gel intravaginally daily x 5 days OR 500 mg PO bid x 7 days. 2) clindamycin 5 g of 2% cream intravaginally qhs x 7 days. In pregnancy: 1) metronidazole 500 mg PO bid x 7 days OR 250 mg PO tid x 7 days. 2) clindamycin 300 mg PO bid x 7 days.

Candidal vaginitis: 1) intravaginal clotrimazole, miconazole, terconazole, nystatin, tioconazole, or butoconazole. 2) fluconazole 150 mg PO single dose.

Chlamydia: First line either azithromycin 1 g PO single dose or doxycycline 100 mg PO bid x 7 days. Second line fluoroquinolones or erythromycin. In pregnancy: 1) azithromycin 1 g PO single dose. 2) amoxicillin 500 mg PO tid x 7 days. Repeat NAAT‡ 3 weeks after treatment.

Epididymitis: 1) ceftriaxone 250 mg IM single dose + doxycycline 100 mg PO bid x 10 days. 2) ofloxacin 300 mg PO bid or levofloxacin 500 mg PO daily x 10 days if enteric organisms suspected, or cephalosporin/doxycycline allergic.

Gonorrhea: Single dose of: 1) ceftriaxone 125 mg IM 2) cefixime 400 mg PO 3) ciprofloxacin 500 mg PO† 4) ofloxacin 400 mg PO† or 5) levofloxacin 250 mg PO.† Treat chlamydia empirically. Cephalosporin desensitization advised for cephalosporin-allergic patients who cannot take fluoroquinolones (e.g. pregnant women). Consider azithromycin 2 g PO single dose for uncomplicated gonorrhea, but no efficacy/safety data for this regimen in pregnant women.

Herpes simplex (genital, first episode): 1) acyclovir 400 mg PO tid x 7-10 days. 2) famciclovir 250 PO tid x 7-10 days. 3) valacyclovir 1 g PO bid x 7-10 days.

Herpes simplex (genital, recurrent): 1) acyclovir 400 mg PO tid x 5 days. 2) acyclovir 800 mg PO tid x 2 days or bid x 5 days. 3) famciclovir 125 mg PO bid x 5 days. 4) famciclovir 1000 mg PO bid x 1 day. 5) valacyclovir 500 mg PO bid x 3 days. 6) valacyclovir 1 g PO daily x 5 days.

Herpes simplex (suppressive therapy): 1) acyclovir 400mg PO bid. 2) famciclovir 250 mg PO bid. 3) valacyclovir 500-1000 mg PO daily.

Pelvic inflammatory disease (PID), outpatient treatment: 1) ceftriaxone 250 mg IM single dose + doxycycline 100 mg PO bid +/- metronidazole 500 mg PO bid x 14 days. 2) ofloxacin 400 mg PO bid† / levofloxacin 500 mg PO daily† +/- metronidazole 500 mg PO bid x 14 days.

Sexual assault prophylaxis: ceftriaxone 125 mg IM single dose + metronidazole 2 g PO single dose + azithromycin 1 g PO single dose/doxycycline 100 mg PO bid x 7 days. Consider giving antiemetic.

Syphilis (primary and secondary): 1) benzathine penicillin 2.4 million units IM single dose. 2) doxycycline 100 mg PO bid x 2 weeks if penicillin allergic.

Trichomoniasis: metronidazole (can use in pregnancy) or tinidazole, each 2 g PO single dose.

Urethritis, Cervicitis: Test for Chlamydia and gonorrhea with NAAT‡. Treat based on test results or treat presumptively if high-risk of infection (Chlamydia: age <=25 y, new/ multiple sex partners, or unprotected sex; gonorrhea: population prevalence >5%), esp. if NAAT‡ unavailable or patient unlikely to return for follow-up.

Urethritis (persistent/recurrent): 1) metronidazole/ tinidazole 2 g PO single dose + azithromycin 2 g PO single dose (if not used in first episode).

---

* MMWR 2006;55:RR-11 or http://www.cdc.gov/STD/treatment/. Treat sexual partners for all except herpes, candida, and bacterial vaginosis.

† Due to high resistance rates, quinolones not recommended if infection acquired in Hawaii or California, recent foreign travel by patient/ partner, or in men who have sex with men. See health department or www.cdc.gov/std/gisp for current resistance info.

‡NAAT = nucleic acid amplification test.

CEPHALOSPORINS – GENERAL ANTIMICROBIAL SPECTRUM

1st generation: gram positive (including Staph aureus); basic gram neg. coverage
2nd generation: diminished Staph aureus, improved gram negative coverage compared to 1st generation; some with anaerobic coverage
3rd generation: further diminished Staph aureus, further improved gram negative coverage compared to 1st & 2nd generation; some with Pseudomonal coverage & diminished gram positive coverage
4th generation: same as 3rd generation plus coverage against Pseudomonas

cefoperazone (**Cefobid**): Usual dose 2-4 g/day IM/IV given q12h. Maximum dose: 6-12 g/day IV given q6-12 h. Possible clotting impairment. ▶Bile/K ♀B ▶? $$$$$

cefotaxime (**Claforan**): Usual dose: 1-2 g IM/IV q6-8h. Peds: 50-180 mg/kg/day IM/IV divided q4-6h. AAP dose for pneumococcal meningitis: 225-300 mg/kg/day IV divided q6-8h. ▶KL ♀B ▶+ $$$$$

cefpodoxime (**Vantin**): 100-400 mg PO bid. Peds: 10 mg/kg/day divided bid. [Generic/Trade: Tabs 100,200 mg. Trade: susp 50 & 100 mg/5 mL.] ▶K ♀B ▶? $$$$

ceftazidime (**Ceptaz, Fortaz, Tazicef**): 1 g IM/IV or 2 g IV q8-12h. Peds: 30-50 mg/kg IV q8h. ▶K ♀B ▶+ $$$$$

ceftibuten (**Cedax**): 400 mg PO daily. Peds: 9 mg/kg up to 400 mg PO daily. [Trade only: Cap 400 mg, susp 90 mg/5 mL.] ▶K ♀B ▶? $$$$

ceftizoxime (**Cefizox**): 1-2 g IV q8-12h. Peds: 50 mg/kg/dose IV q6-8h. ▶K ♀B ▶? $$$$$

ceftriaxone (**Rocephin**): 1-2 g IM/IV q24h. Meningitis: 2 g IV q12h. Gonorrhea: single dose 125 mg IM (250 mg if PID). Peds: 50-75 mg/kg/day up to 2 g divided q12-24h. Meningitis: 100 mg/kg/day up to 4 g/day. Otitis media: 50 mg/kg up to 1 g IM single dose. Dilute in 1% lidocaine for IM. ▶K/Bile ♀B ▶+ $$$$$

Cephalosporins - 4th Generation

cefepime (**Maxipime**): 0.5-2 g IM/IV q12h. Peds: 50 mg/kg IV q8-12h. ▶K ♀B ▶? $$$$$

Ketolides

telithromycin (**Ketek**): 800 mg PO daily x 5 days for acute exacerbation of chronic bronchitis or acute sinusitis, x 7 days for community-acquired pneumonia. [Trade only: 300,400 mg tabs. Ketek Pak: #10, 400 mg tabs.] ▶LK ♀C ▶? $$$

Macrolides

azithromycin (**Zithromax, Zmax**): 500 mg IV daily. PO: 10 mg/kg up to 500 mg on day 1, then 5 mg/kg up to 250 mg daily to complete 5 days. Group A strep pharyngitis (2nd-line to penicillin): 12 mg/kg up to 500 mg PO daily x 5 d. Short regimens for peds otitis media (30 mg/kg PO single dose or 10 mg/kg PO daily x 3 days) and sinusitis (10 mg/kg PO daily x 3 days). Short regimen for adult acute sinusitis or exacerbation of chronic bronchitis: 500 mg PO daily x 3 days. Chlamydia, chancroid: 1 g PO single dose. Prevention of disseminated Mycobacterium avium complex disease: 1200 mg PO q week. Acute sinusitis in children: 10 mg/kg PO daily x 3 days. [Generic/Trade: Tab 250, 500, 600 mg, Susp 100 & 200/5 mL. Trade only: Packet 1000 mg. Z-Pak: #6, 250 mg tab. Tri-Pak: #3, 500 mg tab. Zmax extended release oral susp: 2 g in 60 mL single dose bottle.] ▶L ♀B ▶? $$

clarithromycin (**Biaxin, Biaxin XL**): 250-500 mg PO bid. Peds: 7.5 mg/kg PO bid. H pylori: See table in GI section. See table for prophylaxis of bacterial endocarditis. Mycobacterium avium complex disease prevention: 7.5 mg/kg up to 500 mg PO bid. *Biaxin XL*: 1000 mg PO daily with food. [Generic/Trade: Tab 250, 500 mg. Extended release 500 mg. Trade only: susp 125, 250 mg/5 mL. Biaxin XL-Pak: #14, 500 mg tabs. Generic only: Extended release 1000 mg.] ▶KL ♀C ▶? $$$

erythromycin base (*Eryc, E-mycin, Ery-Tab, ✚Erybid, Erythromid, P.C.E.*): 250-500 mg PO qid, 333 mg PO tid, or 500 mg PO bid. [Generic/Trade: Tab 250, 333, 500 mg, delayed-release cap 250.] ▶L ♀B ▶+ $

erythromycin ethyl succinate (*EES, Eryped*): 400 mg PO qid. Peds: 30-50 mg/kg/day divided qid. [Generic/Trade: Tab 400 mg, susp 200 & 400 mg/5 mL. Trade only (EryPed): Susp 100 mg/2.5mL.] ▶L ♀B ▶+ $

erythromycin lactobionate (*✚Erythrocin IV*): 15-20 mg/kg/day (max 4g) IV divided q6h. Peds: 15-50 mg/kg/day IV divided q6h. ▶L ♀B ▶+ $$$$

Pediazole (erythromycin ethyl succinate + sulfisoxazole): 50 mg/kg/day (based on EES dose) PO divided tid-qid. [Generic/Trade: Susp, erythromycin ethyl succinate 200 mg + sulfisoxazole 600 mg/5 mL.] ▶KL ♀C ▶- $$

Penicillins - 1st generation - Natural

benzathine penicillin (*Bicillin L-A, ✚Megacillin*): 1.2 million units IM. Peds <27 kg 0.3-0.6 MU IM, ≥27 kg 0.9 MU IM. Doses last 2-4 wks. [Trade only: for IM use, 600,000 units/mL; 1, 2, and 4 mL syringes.] ▶K ♀B ▶? $$

benzylpenicilloyl polylysine (*Pre-Pen*): Skin test for penicillin allergy: 1 drop in needle scratch, then 0.01-0.02 mL intradermally if no reaction. ▶K ♀? ▶? $

Bicillin C-R (procaine penicillin + benzathine penicillin): For IM use. Not for treatment of syphilis. [Trade only: for IM use 300/300 thousand units procaine/benzathine penicillin; 1, 2, and 4 mL syringes.] ▶K ♀B ▶? $$$

penicillin G: Pneumococcal pneumonia & severe infections: 250,000-400,000 units/kg/day (8-12 million units in adult) IV divided q4-6h. Pneumococcal meningitis: 250,000 units/kg/day (24 million units in an adult) IV divided q2-4h. ▶K ♀B ▶? $$$

PROPHYLAXIS FOR BACTERIAL ENDOCARDITIS*

For dental, oral, respiratory tract, or esophageal procedures	
Standard regimen	amoxicillin[1] 2 g PO 1h before procedure
Unable to take oral meds	ampicillin[1] 2 g IM/IV within 30 minutes before procedure
Allergic to penicillin	clindamycin[2] 600 mg PO; or cephalexin[1] or cefadroxil[1] 2 g PO; or azithromycin[3] or clarithromycin[3] 500 mg PO 1h before procedure
Allergic to penicillin and unable to take oral meds	clindamycin[2] 600 mg IV; or cefazolin[4] 1 g IM/IV within 30 minutes before procedure
For genitourinary and gastrointestinal (excluding esophageal) procedures	
High-risk patients	ampicillin[1] 2 g IM/IV plus gentamicin 1.5 mg/kg (max 120 mg) within 30 min of starting procedure; 6h later ampicillin[1] 1 g IM/IV or amoxicillin[1] 1 g PO.
High-risk patients allergic to ampicillin	vancomycin[2] 1 g IV over 1-2h plus gentamicin 1.5 mg/kg IV/IM (max 120 mg) complete within 30 minutes of starting procedure
Moderate-risk patients	amoxicillin[1] 2 g PO or ampicillin[1] 2 g IM/IV within 30 minutes of starting procedure
Moderate-risk patients allergic to ampicillin	vancomycin[2] 1 g IV over 1-2h complete within 30 minutes of starting procedure

*JAMA 1997; 277:1794-1801 or http://www.americanheart.org
Footnotes for pediatric doses: 1 = 50 mg/kg; 2 = 20 mg/kg; 3 = 15 mg/kg; 4 = 25 mg/kg.
Total pediatric dose should not exceed adult dose.

PENICILLINS - GENERAL ANTIMICROBIAL SPECTRUM

1st generation: Most streptococci; oral anaerobic coverage
2nd generation: Most streptococci; Staph aureus
3rd generation: Most streptococci; basic gram negative coverage
4th generation: Pseudomonas

penicillin V (**Pen-Vee K, Veetids, ✚PVF-K, Nadopen-V**): Adults: 250-500 mg PO qid. Peds: 25-50 mg/kg/day divided bid-qid. AHA doses for pharyngitis: 250 mg (peds) or 500 mg (adults) PO bid-tid x 10 days. [Generic/Trade: Tabs 250,500 mg, oral soln 125 & 250 mg/5 mL.] ▶K ♀B ▶? $

procaine penicillin (**Wycillin**): 0.6-1.0 million units IM daily (peak 4h, lasts 24h). ▶K ♀B ▶? $$$$

Penicillins - 2nd generation - Penicillinase-Resistant

dicloxacillin (**Dynapen**): 250-500 mg PO qid. Peds: 12.5-25 mg/kg/day divided qid. [Generic: Caps 250,500 mg. Trade only: Susp 62.5 mg/5 mL.] ▶KL ♀B ▶? $$

nafcillin: 1-2 g IM/IV q4h. Peds: 50-200 mg/kg/day divided q4-6h. ▶L ♀B ▶? $$$$$

oxacillin (**Bactocill**): 1-2 g IM/IV q4-6h. Peds 150-200 mg/kg/day IM/IV divided q4-6h. ▶KL ♀B ▶? $$$$$

Penicillins - 3rd generation - Aminopenicillins

amoxicillin (**Amoxil, DisperMox, Polymox, Trimox, ✚Novamoxin**): 250-500 mg PO tid, or 500-875 mg PO bid. Acute sinusitis with antibiotic use in past month &/ or drug-resistant S pneumoniae rate >30%: 3-3.5 g/day PO. High-dose for community-acquired pneumonia: 1 g PO tid. Peds AAP otitis media: 80-90 mg/kg/day divided bid-tid. AAP recommends 5-7 days of therapy for older (≥6 yo) children with non-severe otitis media, and 10 days for younger children and those with severe disease. Peds non-otitis: 40 mg/kg/day PO divided tid or 45 mg/kg/day divided bid. [Generic/Trade: Caps 250,500 mg, tabs 500,875 mg, chews 125, 200, 250,400mg, susp 125, 250 mg/5 mL, susp 200, 400 mg/5 mL. Trade: Infant drops 50 mg/mL (Amoxil). DisperMox 200,400,600 mg tabs for oral susp.] ▶K ♀B ▶+ $

amoxicillin-clavulanate (**Augmentin, Augmentin ES-600, Augmentin XR, ✚Clavulin**): 500-875 mg PO bid or 250-500 mg bid. Peds AAP otitis media: *Augmentin XR*: 2 tabs PO q12h with meals. Peds AAP otitis media: *Augmentin ES* 90 mg/kg/day divided bid. AAP recommends 5-7 days of therapy for older (≥6 yo) children with non-severe otitis media, and 10 days for younger children and those with severe disease. Peds: 45 mg/kg/day PO divided bid or 40 mg/kg/day divided tid for otitis, sinusitis, pneumonia; 25 mg/kg/day PO divided bid or 20 mg/kg/day divided tid for less severe infections. [Generic/Trade: (amoxicillin + clavulanate) Tabs 250+125, 500+125, 875+ 125 mg, chewables & susp 200+28.5, 400+57 mg per tab or 5 mL, (ES) susp 600 +42.9 mg/5mL Trade only: Chewables and susp 125+31.25, 250+62.5 mg per tab or 5 mL. Extended-release tabs (Augmentin XR) 1000+62.5 mg.] ▶K ♀B ▶? $$$

ampicillin (**Principen, ✚Penbritin**): Usual dose: 1-2 g IV q4-6h. Sepsis, meningitis: 150-200 mg/kg/day IV divided q3-4h. Peds: 50-400 mg/kg/day IM/IV divided q4-6h. [Generic/Trade: Caps 250,500 mg, susp 125 & 250 mg/5 mL.] ▶K ♀B ▶? $ PO $$$$$ IV

ampicillin-sulbactam (**Unasyn**): 1.5-3 g IM/IV q6h. Peds: 100-400 mg/kg/day of ampicillin divided q6h. ▶K ♀B ▶? $$$$$

pivampicillin (**Pondocillin**): Canada only. Adults: 500-1000 mg PO bid. Infants, 3-12 mo: 40-60 mg/kg/day PO divided bid. Peds, 1-10 yo: 25-35 mg/kg/day PO divided bid up to 525 mg PO bid. [Trade only: Tabs 500 mg (377 mg ampicillin), # 20, oral susp 35 mg/mL (26 mg ampicillin), 100,150,200 mL bottles.] ▶K ♀? ▶? $

Penicillins - 4th generation - Extended Spectrum

carbenicillin (*Geocillin*): UTI: 382-764 mg PO qid. Prostatitis: 764 mg PO qid. [Trade only: Tab 382 mg.] ▶K ♀B ▶? $$$$$

piperacillin: 3-4 g IM/IV q4-6h. ▶K/Bile ♀B ▶? $$$$$

piperacillin-tazobactam (*Zosyn*, ♥*Tazocin*): 3.375-4.5 g IV q6h. Peds: 240 mg/kg/day of piperacillin IV divided q8h. ▶K ♀B ▶? $$$$$

ticarcillin (*Ticar*): 3-4 g IM/IV q4-6h. Peds: 200-300 mg/kg/d divided q4-6h. ▶K ♀B ▶+ $$$$$

ticarcillin-clavulanate (*Timentin*): 3.1 g IV q4-6h. Peds: 50 mg/kg up to 3.1 g IV q4-6h. ▶K ♀B ▶? $$$$$

Quinolones - 1st Generation

nalidixic acid (*NegGram*): 1 g PO qid. [Trade: Tabs 0.25,0.5,1 g.] ▶KL ♀C ▶? $$$$

Quinolones - 2nd Generation

ciprofloxacin (*Cipro, Cipro XR, ProQuin XR*): 200-400 mg IV q8-12h. 250-750 mg PO bid. Simple UTI: 250 mg bid x 3d or Cipro XR/Proquin XR 500 mg PO daily x 3d. Give Proquin XR with main meal of day. Cipro XR for pyelonephritis or complicated UTI: 1000 mg PO daily x 7-14 days. Gonorrhea: 500 mg PO single dose (not for infections acquired in California, Hawaii, Asia, or Pacific islands, or in men who have sex with men). [Generic/Trade: Susp 250 & 500 mg/5 mL. Tabs 100, 250, 500, 750 mg. Trade only: Extended release tabs (Cipro XR & Pro-Quin XR), 1000 mg (Cipro XR only).] ▶LK ♀C but teratogenicity unlikely ▶?+ $$$$

lomefloxacin (*Maxaquin*): 400 mg PO daily. Take at night. Photosensitivity. [Trade only: Tabs 400 mg.] ▶LK ♀C ▶? $$$

norfloxacin (*Noroxin*): Simple UTI: 400 mg PO bid x 3 days. [Trade only: Tabs 400 mg.] ▶LK ♀C ▶? $$$

ofloxacin (*Floxin*): 200-400 mg IV/PO q12h. [Generic/Trade: Tabs 200,300,400 mg.] ▶LK ♀C ▶?+ $$$

Quinolones - 3rd Generation

levofloxacin (*Levaquin*): 250-750 mg PO/IV daily. [Trade only: Tabs 250,500,750 mg. Trade only: oral soln 25 mg/mL. Leva-Pak: #5, 750 mg tabs.] ▶KL ♀C ▶? $$$

Quinolones - 4th Generation

gatifloxacin (*Tequin*): 400 mg IV/PO daily. Simple UTI: 400 mg PO single dose or 200 mg PO daily x 3 d. Gonorrhea: 400 mg PO single dose (not for infections acquired in California, Hawaii, Asia, or Pacific islands, or in men who have sex with men). [Trade only: Tabs 200,400 mg. Susp 200 mg/5 mL. Teq-Paq: #5, 400 mg tabs.] ▶K ♀C ▶- $$$

gemifloxacin (*Factive*): 320 mg PO daily x 5-7 days. [Trade only: Tabs 320 mg.] ▶Feces, K ♀C ▶- $$$

moxifloxacin (*Avelox*): 400 mg PO/IV daily x 5 days (chronic bronchitis exacerbation), 5-14 days (complicated intra-abdominal infection; usually given IV initially), 7 days (uncomplicated skin infections), 10 days (acute sinusitis), 7-14 days (community acquired pneumonia), 7-21 days (complicated skin infections). [Trade only: Tabs 400 mg.] ▶LK ♀C ▶- $$$

QUINOLONES- GENERAL ANTIMICROBIAL SPECTRUM

1st generation: gram negative (excluding Pseudomonas), urinary tract only, no atypicals

2nd generation: gram negative (including Pseudomonas); Staph aureus but not pneumococcus; some atypicals

3rd generation: gram negative (including Pseudomonas); gram positive (including Staph aureus and pneumococcus); expanded atypical coverage

4th generation: same as 3rd generation plus enhanced coverage of pneumococcus, decreased activity vs. Pseudomonas.

Sulfonamides

sulfadiazine: CNS toxoplasmosis in AIDS. 1000 mg (<60 kg) to 1500 mg (≥60 kg) PO qid for acute tx; 500-1000 mg PO qid for secondary prevention. Give with pyrimethamine + leucovorin. [Generic: Tab 500 mg.] ▶K ♀C ▶+ $$$$

trimethoprim-sulfamethoxazole (**Bactrim, Septra, Sulfatrim,** cotrimoxazole): One tab PO bid, double strength (DS, 160 mg/800 mg) or single strength (SS, 80 mg/400 mg). Pneumocystis treatment: 15-20 mg/kg/day (based on TMP) IV divided q6-8h or PO divided tid x 21 days total. Pneumocystis prophylaxis: 1 DS tab PO daily. Peds: 5 mL susp/10 kg (up to 20 mL)/dose PO bid. [Generic/Trade: Tabs 80 mg TMP/400 mg SMX (single strength), 160 mg TMP/800 mg SMX (double strength; DS), susp 40 mg TMP/200 mg SMX per 5 mL. 20 mL susp = 2 SS tabs = 1 DS tab.] ▶K ♀C ▶+ $

Tetracyclines

doxycycline (**Adoxa, Vibramycin, Doryx, Monodox, Oracea, Periostat,** ✚**Doxycin**): 100 mg PO bid on first day, then 50 mg bid or 100 mg daily. 100 mg PO/IV bid for severe infections. Periostat for periodontitis: 20 mg PO bid. Oracea for inflammatory rosacea: 40 mg PO once every morning on empty stomach. [Generic/Trade: Tabs 75,100 mg, caps 50,100 mg. Trade only: Susp 25 mg/5 mL, syrup 50 mg/5 mL, Periostat: Tabs 20 mg. Oracea: Caps 40 mg.] ▶LK ♀D ▶?+ $

minocycline (**Minocin, Dynacin, Solodyn**): 200 mg IV/PO initially, then 100 mg q12h. Solodyn for inflammatory acne, ≥12 yo: Give PO once daily at dose of 45 mg for 99-131 lb, 90 mg for 132-199 lb, 135 mg for 200-300 lb. [Generic/Trade: Caps, tabs 50,75,100 mg. Trade only (Solodyn): Extended release tabs 45, 90, 145 mg.] ▶LK ♀D ▶?+ $$$

tetracycline (**Sumycin**): 250-500 mg PO qid. [Generic/Trade: Caps 250,500 mg. Trade only: Tabs 250, 500 mg, susp 125 mg/5 mL.] ▶LK ♀D ▶?+ $

Other Antimicrobials

aztreonam (**Azactam**): 0.5-2 g IM/IV q6-12h. Peds: 30 mg/kg q6-8h. ▶K ♀B ▶+ $$$$$

chloramphenicol (**Chloromycetin**): 50-100 mg/kg/day IV divided q6h. Aplastic anemia. ▶LK ♀C ▶- $$$$$

clindamycin (**Cleocin,** ✚**Dalacin C**): 600-900 mg IV q8h. Each IM injection should be ≤600 mg. 150-450 mg PO qid. Peds: 20-40 mg/kg/day IV divided q6-8h or 8-25 mg/kg/day susp PO divided tid-qid. [Generic/Trade: Cap 75,150,300 mg. Trade only: Oral soln 75 mg/5 mL.] ▶L ♀B ▶?+ $$$

daptomycin (**Cubicin, Cidecin**): Complicated skin infections: 4 mg/kg IV daily x 7-14 days. S aureus bacteremia: 6 mg/kg IV daily x ≥2-6 weeks. Infuse over 30 min. ▶K ♀B ▶? $$$$$

drotrecogin (**Xigris**): To reduce mortality in sepsis: 24 mcg/kg/h IV x 96 h. ▶Plasma ♀C ▶? $$$$$

fosfomycin (**Monurol**): Simple UTI: One 3 g packet PO single-dose. [Trade only: 3 g packet of granules.] ▸K ♀B ▸? $$

linezolid (**Zyvox**, ♣**Zyvoxam**): 600-400 mg IV/PO q12 h. Infuse over 30-120 minutes. Peds: 10 mg/kg IV/PO q8h. Uncomplicated skin infections: 10 mg/kg PO q8h if <5 yo, q12h if 5-11 yo. Myelosuppression. MAO inhibitor. [Trade only: Tabs 600 mg, susp 100 mg/5 mL.] ▸Oxidation/K ♀C ▸? $$$$$

metronidazole (**Flagyl**, ♣**Florazole ER, Trikacide, Nidazol**): Bacterial vaginosis: 500 mg PO bid x 7 days or Flagyl ER 750 mg PO daily x 7 days. H pylori: See table in GI section. Anaerobic bacterial infections: Load 1 g or 15 mg/kg IV, then 500 mg or 7.5 mg/kg IV/PO q6-8h, each IV dose over 1h (not to exceed 4 g/day). Peds: 7.5 mg/kg IV q6h. C difficile diarrhea: 500 mg (10-15 mg/kg/dose for peds) PO tid. [Generic/Trade: Tabs 250,500 mg, ER tabs 750 mg, Caps 375 mg.] ▸KL ♀B ▸?- $

nitrofurantoin (**Furadantin, Macrodantin, Macrobid**): 50-100 mg PO qid. Peds: 5-7 mg/kg/day divided qid. Sustained release: 100 mg PO bid. [Macrodantin: Caps 25,50,100 mg. Generic: Caps 50,100 mg. Furadantin: Susp 25 mg/5 mL. Macrobid: Caps 100 mg.] ▸KL ♀B ▸+? $$$

rifampin (**Rimactane, Rifadin**, ♣**Rofact**): Neisseria meningitidis carriers: 600 mg PO bid x 2 days. Peds: age ≥1 mo, 10 mg/kg up to 600 mg PO bid x 2 days. Age <1 mo, 5 mg/kg PO bid x 2 days. IV & PO doses are the same. [Generic/Trade: Caps 150,300 mg. Pharmacists can make oral suspension.] ▸L ♀C ▸+ $$$$

rifaximin (**Xifaxan**): Travelers diarrhea: 200 mg PO tid x 3 days. [Trade only: Tab 200 mg.] ▸Feces, no GI absorption ♀C ▸? $$$

Synercid (quinupristin + dalfopristin): 7.5 mg/kg IV q8-12 h, each dose over 1 h. Not active against E. faecalis. ▸Bile ♀B ▸? $$$$$

tigecycline (**Tygacil**): Complicated skin or intra-abdominal infections: 100 mg IV first dose, then 50 mg IV q12h. Infuse over 30-60 minutes. ▸Bile, K ♀D ▸?+ $$$$$

trimethoprim (**Primsol**, ♣**Proloprim**): 100 mg PO bid or 200 mg PO daily. [Generic: Tabs 100,200 mg. Primsol: Oral soln 50 mg/5 mL.] ▸K ♀C ▸- $

vancomycin (**Vancocin**): 1g IV q12h, each dose over 1h. Peds: 10-15 mg/kg IV q6h. Clostridium difficile diarrhea: 40-50 mg/kg/day up to 500 mg/day divided qid x 7-10 days. IV administration ineffective for this indication. [Trade only: Caps 125,250 mg] ▸K ♀C ▸? $$$$$

CARDIOVASCULAR

ACE Inhibitors

> **NOTE:** See also antihypertensive combinations. Hyperkalemia possible, especially if used concomitantly with other drugs that increase K+ (including K+ containing salt substitutes) and in patients with heart failure, diabetes mellitus, or renal impairment. Monitor closely for hypoglycemia during first month of treatment when combined with insulin or oral antidiabetic agents. ACE inhibitors are contraindicated during pregnancy. Contraindicated with a history of angioedema.

benazepril (**Lotensin**): HTN: Start 10 mg PO daily, usual maintenance dose 20-40 mg PO daily or divided bid, max 80 mg/day. [Generic/Trade: Tabs, non-scored 5,10,20,40 mg.] ▸LK ♀- ▸? $$

captopril (**Capoten**): HTN: Start 25 mg PO bid-tid, usual maintenance dose 25-150 mg PO bid-tid, max 450 mg/day. Heart failure: Start 6.25-12.5 mg PO tid, usual dose 50-100 mg PO tid, max 450 mg/day. Diabetic nephropathy: 25 mg PO tid. [Generic/Trade: Tabs, scored 12.5,25,50,100 mg.] ▸LK ♀- ▸+ $

cilazapril (♣**Inhibace**): Canada only. HTN: 1.25-10 mg PO daily. [Generic/Trade: Scored tabs 1, 2.5, 5 mg.] ▸LK ♀- ▸? $

enalapril (**enalaprilat**, *Vasotec*): HTN: Start 5 mg PO daily, usual maintenance dose 10-40 mg PO daily or divided bid, max 40 mg/day. If oral therapy not possible, can use enalaprilat 1.25 mg IV q6h over 5 minutes, and increase up to 5 mg IV q6h if needed. Renal impairment or concomitant diuretic therapy: Start 2.5 mg PO daily. Heart failure: Start 2.5 mg PO bid, usual 10-20 mg PO bid, max 40 mg/day. [Generic/Trade: Tabs, scored 2.5,5, non-scored 10, 20 mg.] ▶LK ♀- ▶+ $

fosinopril (*Monopril*): HTN: Start 10 mg PO daily, usual maintenance dose 20-40 mg PO daily or divided bid, max 80 mg/day. Heart failure: Start 10 mg PO daily, usual dose 20-40 mg PO daily, max 40 mg/day. [Generic/Trade: Tabs, scored 10, non-scored 20,40 mg.] ▶LK ♀- ▶? $$

lisinopril (*Prinivil*, *Zestril*): HTN: Start 10 mg PO daily, usual maintenance dose 20-40 mg PO daily, max 80 mg/day. Heart failure, acute MI: Start 2.5-5 mg PO daily, usual dose 5-20 mg PO daily, max dose 40 mg. [Generic/Trade: Tabs, non-scored 2.5,10,20,30,40, scored 5 mg.] ▶K ♀- ▶? $$

moexipril (*Univasc*): HTN: Start 7.5 mg PO daily, usual maintenance dose 7.5-30 mg PO daily or divided bid, max 30 mg/day. [Generic/Trade: Tabs, scored 7.5, 15 mg.] ▶LK ♀- ▶? $$

perindopril (*Aceon*, ✚*Coversyl*): HTN: Start 4 mg PO daily, usual maintenance dose 4-8 mg PO daily or divided bid, max 16 mg/day. Reduction of cardiovascular events in stable coronary artery disease: Start 4 mg PO daily x 2 weeks, max 8 mg/day. Elderly (>70 years): 2 mg PO daily x 1 week, 4 mg PO daily x 1 week, max 8 mg/day. [Trade only: Tabs, scored 2,4,8 mg.] ▶K ♀- ▶? $$

quinapril (*Accupril*): HTN: Start 10-20 mg PO daily (start 10 mg/day if elderly), usual maintenance dose 20-80 mg PO daily or divided bid, max 80 mg/day. Heart failure: Start 5 mg PO bid, usual maintenance dose 10-20 mg bid. [Generic/Trade: Tabs, scored 5, non-scored 10, 20, 40 mg.] ▶LK ♀- ▶? $$

ramipril (*Altace*): HTN: 2.5 mg PO daily, usual maintenance dose 2.5-20 mg PO daily or divided bid, max 20 mg/day. Heart failure post-MI: Start 2.5 mg PO bid, usual maintenance dose 5 mg PO bid. Reduce risk of MI, stroke, death from cardiovascular causes: 2.5 mg PO daily x 1 week, then 5 mg daily x 3 weeks, increase as tolerated to max 10 mg/day. [Generic/Trade: Caps 1.25, 2.5, 5, 10 mg.] ▶LK ♀- ▶? $$

trandolapril (*Mavik*): HTN: Start 1 mg PO daily, usual maintenance dose 2-4 mg PO daily or divided bid, max 8 mg/day. Heart failure/post-MI: Start 0.5-1 mg PO daily, usual maintenance dose 4 mg PO daily. [Trade only: Tabs, scored 1, non-scored 2,4 mg.] ▶LK ♀- ▶? $$

ACE INHIBITOR DOSING	Hypertension		Heart Failure		
	Initial	*Max/day*	*Initial*	*Target*	*Max*
benazepril (*Lotensin*)	10 mg qd*	80 mg	-	-	-
captopril (*Capoten*)	25 mg bid/tid	450 mg	6.25-12.5 mg tid	50 mg tid	150 mg tid
enalapril (*Vasotec*)	5 mg qd*	40 mg	2.5 mg bid	10 mg bid	20 mg bid
fosinopril (*Monopril*)	10 mg qd*	80 mg	10 mg qd	20 mg qd	40 mg qd
lisinopril (*Zestril/Prinivil*)	10 mg qd	80 mg	5 mg qd	20 mg qd	40 mg qd
moexipril (*Univasc*)	7.5 mg qd*	30-60 mg	-	-	-
perindopril (*Aceon*)	4 mg qd*	16 mg	-	-	-
quinapril (*Accupril*)	10-20 mg qd*	80 mg	5 mg bid	10 mg bid	20 mg bid
ramipril (*Altace*)	2.5 mg qd*	20 mg	2.5 mg bid	5 mg bid	10 mg bid
trandolapril (*Mavik*)	1-2 mg qd*	8 mg	1 mg qd	4 mg qd	4 mg qd

Data taken from prescribing information.   **May require bid dosing for 24-hour BP control.*

Aldosterone Antagonists

eplerenone (*Inspra*): HTN: Start 50 mg PO daily; max 50 mg bid. Improve survival of stable patients with left ventricular systolic dysfunction (EF ≤40%) and heart failure post-MI: Start 25 mg PO daily; titrate to target dose 50 mg daily within 4 weeks, if tolerated. [Trade only: Tabs non-scored 25, 50 mg] ▶L ♀B ▶? $$$$

spironolactone (*Aldactone*): HTN: 50-100 mg PO daily or divided bid. Edema: 25-200 mg/day. Hypokalemia: 50-100 mg PO daily. Primary hyperaldosteronism, maintenance: 100-400 mg/day PO. Cirrhotic ascites: Start 100 mg once daily or in divided doses. Maintenance 25-200 mg/day. [Generic/Trade: Tabs, non-scored 25; scored 50,100 mg.] ▶LK ♀D ▶+ $

Angiotensin Receptor Blockers (ARBs) See also antihypertensive combinations.

candesartan (*Atacand*): HTN: Start 16 mg PO daily, maximum 32 mg/day. Reduce cardiovascular death and hospitalizations from heart failure (NYHA II-IV and ejection fraction ≤40%): Start 4 mg PO daily, maximum 32 mg/day; has added effect when used with ACE inhibitor. [Trade only: Tabs, non-scored 4,8,16,32 mg.] ▶K ♀C (1st trimester) D (2nd & 3rd) ▶? $$$

eprosartan (*Teveten*): HTN: Start 600 mg PO daily, maximum 800 mg/day given daily or divided bid. [Trade only: Tabs non-scored 400, 600 mg.] ▶Fecal excretion ♀C (1st trimester) D (2nd & 3rd) ▶? $$

irbesartan (*Avapro*): HTN: Start 150 mg PO daily, max 300 mg/day. Type 2 diabetic nephropathy: Start 150 mg PO daily, target dose 300 mg daily. [Trade only: Tabs, non-scored 75,150,300 mg.] ▶L ♀C (1st trimester) D (2nd & 3rd) ▶? $$$

losartan (*Cozaar*): HTN: Start 50 mg PO daily, max 100 mg/day given daily or divided bid. Stroke risk reduction in patients with HTN & left ventricular hypertrophy (may not occur in Blacks): Start 50 mg PO daily. If need more BP reduction: add HCTZ 12.5 mg PO daily; then increase losartan to 100 mg/day, then increase HCTZ to 25 mg/day. Type 2 diabetic nephropathy: Start 50 mg PO daily, target dose 100 mg daily. [Trade only: Tabs, non-scored 25,50,100 mg.] ▶L ♀C (1st trimester) D (2nd & 3rd) ▶? $$$

olmesartan (*Benicar*): HTN: Start 20 mg PO daily, max 40 mg/day. [Trade only: Tabs, non-scored 5,20,40 mg.] ▶K ♀C (1st trimester) D (2nd & 3rd) ▶? $$

telmisartan (*Micardis*): HTN: Start 40 mg PO daily, max 80 mg/day. [Trade only: Tabs, non-scored 20,40,80 mg.] ▶L ♀C (1st trimester) D (2nd & 3rd) ▶? $$$

valsartan (*Diovan*): HTN: Start 80-160 mg PO daily, max 320 mg/day. Heart failure: Start 40 mg PO bid, target dose 160 mg bid; provides no added effect when used with adequate dose of ACE inhibitor. Reduce mortality/morbidity post-MI with left ventricular systolic dysfunction/failure: Start 20 mg PO bid, target dose 160 mg bid. [Trade only: Tabs, scored 40, nonscored 80, 160, 320 mg.] ▶L ♀C (1st trimester) D (2nd & 3rd) ▶? $$$

Antiadrenergic Agents

clonidine (*Catapres, Catapres-TTS*, ✚*Dixarit*): HTN: Start 0.1 mg PO bid, usual maintenance dose 0.2 to 1.2 mg/day divided bid-tid, max 2.4 mg/day. Rebound HTN with abrupt discontinuation, especially at doses ≥0.8 mg/d. Transdermal (Catapres-TTS): Start 0.1 mg/24 hour patch q week, titrate to desired effect, max effective dose 0.6 mg/24 hour (two, 0.3 mg/24 hour patches). [Generic/Trade: Tabs, non-scored 0.1, 0.2, 0.3 mg. Trade only: transdermal weekly patch 0.1 mg/day (TTS-1), 0.2 mg/day (TTS-2), 0.3 mg/day (TTS-3).] ▶LK ♀C ▶? $

doxazosin (*Cardura, Cardura XL*): HTN: Start 1 mg PO qhs, max 16 mg/day. Take first dose at bedtime to minimize orthostatic hypotension. BPH: see urology section. Extended release form is not FDA approved for HTN. [Generic/Trade: Tabs, scored 1,2,4,8 mg. Trade only: XL tabs 4,8 mg.] ▶L ♀C ▶? $

guanfacine (**Tenex**): Start 1 mg PO qhs, increase to 2-3 mg qhs if needed after 3-4 weeks, max 3 mg/day. ADHD in children: Start 0.5 mg PO daily, titrate by 0.5 mg q3-4 days as tolerated to 0.5 mg PO tid. [Generic/Trade: Tabs, non-scored 1,2 mg.] ▶K ♀B ▶? $

methyldopa (**Aldomet**): HTN: Start 250 mg PO bid-tid, maximum 3000 mg/day. May cause hemolytic anemia. [Generic/Trade: Tabs, non-scored 500 mg. Generic: Tabs, non-scored 125, 250 mg.] ▶LK ♀B ▶+ $

prazosin (**Minipress**): HTN: Start 1 mg PO bid-tid, max 40 mg/day. Take first dose at bedtime to minimize orthostatic hypotension. [Generic/Trade: Caps 1,2,5 mg.] ▶L ♀C ▶? $

terazosin (**Hytrin**): HTN: Start 1 mg PO qhs, usual effective dose 1-5 mg PO daily or divided bid, max 20 mg/day. Take first dose at bedtime to minimize orthostatic hypotension. [Generic (Caps, Tabs)/Trade (Caps): 1,2,5,10 mg.] ▶LK ♀C ▶? $$

Anti-Dysrhythmics / Cardiac Arrest

adenosine (**Adenocard**): PSVT conversion (not A-fib): Adult and peds ≥50 kg: 6 mg rapid IV & flush, preferably through a central line. If no response after 1-2 mins then 12 mg. A third dose of 12 mg may be given prn. Peds <50 kg: initial dose 50-100 mcg/kg, subsequent doses 100-200 mcg/kg q1-2 min prn up to a max single dose of 300 mcg/kg or 12 mg. Half-life is <10 seconds. Give doses by rapid IV push followed by normal saline flush. Need higher dose if on theophylline or caffeine, lower dose if on dipyridamole or carbamazepine. ▶Plasma ♀C ▶? $$$

amiodarone (**Cordarone, Pacerone**): Life-threatening ventricular arrhythmia without cardiac arrest: Load 150 mg IV over 10 min, then 1 mg/min x 6h, then 0.5 mg/min x 18h. Mix in D5W. Oral loading dose 800-1600 mg PO daily for 1-3 weeks, reduce to 400-800 mg PO daily for 1 month when arrhythmia is controlled, reduce to lowest effective dose thereafter, usually 200-400 mg PO daily. Photosensitivity with oral therapy. Pulmonary & hepatic toxicity. Hypo or hyperthyroidism possible. Co-administration of fluoroquinolones, macrolides, or azoles may prolong QTc. May increase digoxin levels; discontinue digoxin or decrease dose by 50%. May increase INR with warfarin by 100%; decrease warfarin dose by 33-50%. Do not use with grapefruit juice. Caution with simvastatin >20 mg/day or lovastatin >40 mg/day; may cause myopathy and rhabdomyolysis. Caution with beta blockers and calcium channel blockers. IV therapy may cause hypotension. Contraindicated with marked sinus bradycardia and second or third degree heart block in the absence of a functioning pacemaker. [Generic: Tabs, scored 300 mg. Generic/Trade: Tabs, scored 200; unscored 100,400 mg.] ▶L ♀D ▶- $$$$

atropine: Bradyarrhythmia/CPR: 0.5-1.0 mg IV q3-5 min to max 0.04 mg/kg. Peds: 0.02 mg/kg/dose; minimum single dose, 0.1 mg; max cumulative dose, 1 mg. ▶K ♀C ▶- $

bicarbonate: Severe acidosis: 1 mEq/kg IV up to 50-100 mEq/dose. ▶K ♀C ▶? $

digoxin (**Lanoxin, Lanoxicaps, Digitek**): Systolic heart failure: 0.125-0.25 mg PO daily; impaired renal function: 0.0625-0.125 mg PO daily. Rapid A-fib: Load 0.5 mg IV, then 0.25 mg IV q6h x 2 doses, maintenance 0.125-0.375 mg IV/PO daily. [Generic/Trade: Tabs, scored (Lanoxin, Digitek) 0.125,0.25 mg; elixir 0.05 mg/mL. Trade only: Caps (Lanoxicaps), 0.05,0.1,0.2 mg.] ▶KL ♀C ▶+ $

digoxin immune Fab (**Digibind, Digifab**): Digoxin toxicity: 2-20 vials IV, one formula is: Number vials = (serum dig level in ng/mL) x (kg) / 100. ▶K ♀C ▶? $$$$$

disopyramide (**Norpace, Norpace CR, ♣Rythmodan, Rythmodan-LA**): Rarely indicated, consult cardiologist. Ventricular arrhythmia: 400-800 mg PO daily in divided doses (immediate-release, q6h or extended-release, q12h). Proarrhythmic. [Generic/Trade: Caps, immediate-release, 100,150 mg; extended-release 150 mg. Trade only: extended-release 100 mg.] ▶KL ♀C ▶+ $$$$

flecainide (**Tambocor**): Proarrhythmic. Prevention of paroxysmal atrial fib/flutter or PSVT, with symptoms & no structural heart disease: Start 50 mg PO q12h, may increase by 50 mg bid q4 days, max 300 mg/day. Use with AV nodal slowing agent (beta blocker, verapamil, diltiazem) to minimize risk of 1:1 atrial flutter. Life-threatening ventricular arrhythmias without structural heart disease: Start 100 mg PO q12h, may increase by 50 mg bid q 4 days, max 400 mg/day. With severe renal impairment (CrCl<35 mL/min): Start 50 mg PO bid. [Generic/Trade: Tabs, non-scored 50, scored 100,150 mg.] ▶K ♀C ▶- $$$$

ibutilide (**Corvert**): Recent onset A-fib/flutter: 0.01 mg/kg up to 1 mg IV over 10 mins, may repeat once if no response after 10 additional minutes. Keep on cardiac monitor ≥4 hours. ▶K ♀C ▶? $$$$$

isoproterenol (**Isuprel**): Refractory bradycardia or third degree AV block: 0.02-0.06 mg IV bolus or infusion 2 mg in 250 mL D5W (8 mcg/mL) at 5 mcg/min. 5 mcg/min = 37 mL/h. Peds: 0.05-2 mcg/kg/min. 10 kg: 0.1 mcg/kg/min = 8 mL/h. ▶LK ♀C ▶? $$

lidocaine (**Xylocaine, Xylocard**): Ventricular arrhythmia: Load 1 mg/kg IV, then 0.5 mg/kg q8-10min as needed to max 3 mg/kg. IV infusion: 4 gm in 500 mL D5W (8 mg/mL) at 1-4 mg/min. Peds: 20-50 mcg/kg/min. ▶LK ♀B ▶? $

mexiletine (**Mexitil**): Rarely indicated, consult cardiologist. Ventricular arrhythmia: Start 200 mg PO q8h with food or antacid, max dose 1,200 mg/day. Proarrhythmic. [Generic/Trade: Caps, 150,200,250 mg.] ▶L ♀C ▶? $$$

procainamide (**Procanbid, Pronestyl**): Ventricular arrhythmia: 500-1250 mg PO q6h or 50 mg/kg/day. Extended-release: 500-1000 mg PO q12h. Load 100 mg IV q10min or 20 mg/min (150 mL) until: 1) QRS widens >50%, 2) dysrhythmia suppressed, 3) hypotension, or 4) total of 17 mg/kg or 1000 mg. Infusion 2g in 250 mL D5W (8 mg/mL) at 2-6 mg/min (15-45 mL/h). Proarrhythmic. [Generic/Trade: Caps, immediate-release 250 mg; tabs, sustained-release, non-scored (Pronestyl SR). Generic only: tabs, sustained-release, non-scored (generic procainamide SR, q6h dosing) 750,1000 mg; caps, immediate-release 500 mg. Trade only: Tabs, immediate-release, non-scored (Pronestyl) 250,375,500 mg, extended-release, non-scored (Procanbid, q12h dosing) 500,1000 mg.] ▶LK ♀C ▶? $$

propafenone (**Rythmol, Rythmol SR**): Proarrhythmic. Prevention of paroxysmal atrial fib/flutter or PSVT, with symptoms & no structural heart disease; or life-threatening ventricular arrhythmias: Start (immediate release) 150 mg PO q8h; may increase after 3-4 days to 225 mg PO q8h; max 900 mg/day. Prolong time to recurrence of symptomatic atrial fib without structural heart disease: 225 mg SR PO q12h, may increase ≥5 days to 325 mg PO q12h, max 425 mg q12h. Use with AV nodal slowing agent (beta blocker, verapamil, diltiazem) to minimize risk of 1:1 atrial flutter. [Generic/Trade: Tabs (immediate release), scored 150,225,300 mg; Trade only: SR, capsules 225,325,425 mg.] ▶L ♀C ▶? $$$$

quinidine (**♣Biquin durules**): Arrhythmia: gluconate, extended-release: 324-648 mg PO q8-12h; sulfate, immediate-release: 200-400 mg PO q6-8h; sulfate, extended-release: 300-600 mg PO q8-12h. Proarrhythmic. [Generic: gluconate: Tabs, extended-release non-scored 324 mg; Generic sulfate: Tabs, scored immediate-release 200,300 mg. Generic sulfate: Tabs, extended-release 300 mg.] ▶LK ♀C ▶? $$-gluconate, $-sulfate

sotalol (**Betapace, Betapace AF, ✦Rylosol, Sotacor, Sotamol**): Ventricular arrhythmia (*Betapace*), A-fib/A-flutter (*Betapace AF*): Start 80 mg PO bid, max 640 mg/d. Proarrhythmic. [Generic/Trade: Tabs, scored 80,120,160,240 mg. Trade only: 80,120,160 mg (*Betapace AF*).] ▶K ♀B ▶ $$$$

SELECTED DRUGS THAT MAY PROLONG THE QT INTERVAL

alfuzosin	droperidol*	isradipine	phenothiazines‡	sotalol*†
amiodarone*†	epirubicin	levofloxacin*	pimozide*†	tacrolimus
arsenic trioxide*	erythromycin*†	lithium	polyethylene glycol	tamoxifen
azithromycin*	felbamate	mefloquine	(PEG-salt solution)§	telithromycin*
chloroquine*	flecainide*	methadone*†	procainamide*	tizanidine
cisapride*†	foscarnet	moexipril/HCTZ	quetiapine‡	tolterodine
clarithromycin*	fosphenytoin	moxifloxacin	quinidine*†	vardenafil
clozapine	gemifloxacin	nicardipine	quinine	venlafaxine
cocaine*	granisetron	octreotide	ranolazine	*Visicol*§
dasatinib	haloperidol*†	ofloxacin	risperidone‡	voriconazole*
disopyramide*†	ibutilide*†	ondansetron	salmeterol	ziprasidone‡
dofetilide*	indapamide*	pentamidine*†	sertraline	

This table may not include all drugs that prolong the QT interval or cause torsades. Risk of drug-induced QT prolongation may be increased in women, elderly, ↓K, ↓Mg, bradycardia, starvation, CHF, and CNS injuries. Hepatorenal dysfunction & drug interactions can ↑ the concentration of QT interval-prolonging drugs. Coadministration of QT interval-prolonging drugs can have additive effects. Avoid these (and other) drugs in congenital prolonged QT syndrome. See www.qtdrugs.org. *Torsades reported in product labeling/case reports. ††Risk in women. ‡QT prolongation: thioridazine>ziprasidone>risperidone, quetiapine, haloperidol. §May be due to electrolyte imbalance.

Anti-Hyperlipidemic Agents - Bile Acid Sequestrants

cholestyramine (**Questran, Questran Light, Prevalite, LoCHOLEST, LoCHO-LEST Light**): Elevated LDL cholesterol: Powder: Start 4 g PO daily-bid before meals, increase up to max 24 g/day. [Generic/Trade: Powder for oral suspension, 4 g cholestyramine resin / 9 g powder (Questran, LoCHOLEST), 4 g cholestyramine resin / 5 g powder (Questran Light), 4 g cholestyramine resin / 5.5 g powder (Prevalite, LoCHOLEST Light).] ▶Not absorbed ♀C ▶+ $

colesevelam (**Welchol**): Elevated LDL cholesterol: 3 Tabs bid with meals or 6 Tabs once daily with a meal, max dose 7 Tabs/day. [Trade only: Tabs, non-scored, 625 mg.] ▶Not absorbed ♀B ▶+ $$$$

colestipol (**Colestid, Colestid Flavored**): Elevated LDL cholesterol: Tabs: Start 2 g PO daily-bid, max 16 g/day. Granules: Start 5 g PO daily-bid, max 30 g/day. [Trade only: Tab 1 g. Generic/Trade: Granules for oral suspension, 5 g / 7.5 g powder.] ▶Not absorbed ♀B ▶+ $$$

Anti-Hyperlipidemic Agents - HMG-CoA Reductase Inhibitors ("Statins")

NOTE: Hepatotoxicity - monitor LFTs initially, approximately 12 weeks after starting therapy, then annually or more frequently if indicated. Evaluate muscle symptoms & creatine kinase before starting therapy. Evaluate muscle symptoms 6-12 weeks after starting/increasing therapy & at each follow-up visit. Obtain creatine kinase when patient complains of muscle soreness, tenderness, weakness, or pain. These factors increase risk of myopathy: advanced age (especially >80, women >men); multisystem disease (eg, chronic renal insufficiency, especially due to diabetes); multiple medications; perioperative periods; alcohol abuse; grapefruit juice (>1 quart/day); specific concomitant medications: fibrates (especially gemfibrozil), nicotinic acid (rare), cyclosporine, erythromycin, clarithromycin, itraconazole, ketoconazole, protease inhibitors, nefazodone, verapamil, amiodarone. Weigh potential risk of combination therapy against potential benefit.

Advicor (lovastatin + niacin): Hyperlipidemia: 1 tab PO qhs with a low-fat snack. Establish dose using extended-release niacin first, or if already on lovastatin substitute combo product with lowest niacin dose. Aspirin or ibuprofen 30 min prior may decrease niacin flushing reaction. [Trade only: Tabs, non-scored extended release niacin/lovastatin 500/20, 750/20, 1000/20, 1000/40 mg.] ▶LK ♀X ▶- $$$

atorvastatin (**Lipitor**): Hyperlipidemia/prevention of cardiovascular events, including type 2 DM: Start 10-40 mg PO daily, max 80 mg/day. [Trade only: Tabs, non-scored 10,20,40,80 mg.] ▶L ♀X ▶- $$$

Caduet (amlodipine + atorvastatin): Simultaneous treatment of HTN and hypercholesterolemia: Establish dose using component drugs first. Dosing interval: daily [Trade only: Tabs, 2.5/10, 2.5/20, 2.5/40, 5/10, 5/20, 5/40, 5/80, 10/10, 10/20, 10/40, 10/80 mg.] ▶L ♀X ▶- $$$$

fluvastatin (**Lescol, Lescol XL**): Hyperlipidemia: Start 20-80 mg PO qhs, max 80 mg daily or divided bid. Post percutaneous coronary intervention: 80 mg of extended release PO daily, max 80 mg daily. [Trade only: Caps, 20,40 mg; tab, extended-release, non-scored 80 mg.] ▶L ♀X ▶- $$$

lovastatin (**Mevacor, Altocor, Altoprev**): Hyperlipidemia/prevention of cardiovascular events: Start 20 mg PO q pm, max 80 mg/day daily or divided bid. [Generic/Trade: Tabs, non-scored 10,20,40 mg. Trade only: Tabs, extended-release (Altocor, Altoprev) 10,20,40,60 mg.] ▶L ♀X ▶- $$$

pravastatin (**Pravachol**): Hyperlipidemia/prevention of cardiovascular events: Start 40 mg PO daily, max 80 mg/day. [Generic/Trade: Tabs, non-scored 10, 20, 40 mg. Trade only: Tabs, non-scored 80 mg.] ▶L ♀X ▶- $$$

rosuvastatin (**Crestor**): Hyperlipidemia: Start 10 mg PO daily, max 40 mg/d. [Trade only: Tabs non-scored 5, 10, 20, 40 mg.] ▶L ♀X ▶- $$$

simvastatin (**Zocor**): Hyperlipidemia: Start 20-40 mg PO q pm, max 80 mg/day. Reduce cardiovascular mortality/events in high risk for coronary heart disease event: Start 40 mg PO q pm, max 80 mg/day. [Generic/Trade: Tabs, non-scored 5, 10, 20, 40, 80 mg.] ▶L ♀X ▶- $$$$

Vytorin (simvastatin + ezetimibe): Hyperlipidemia: Start 10/20 mg PO q pm, max 10/80 mg/day. Start 10/40 if need >55% LDL reduction. [Trade only: Tabs, non-scored ezetimibe/simvastatin 10/10, 10/20, 10/40, 10/80 mg.] ▶L ♀X ▶- $$$

STATINS*

Minimum Dose for 30-40% LDL Reduction	LDL	LFT Monitoring**
atorvastatin 10 mg	-39%	B, 12 wk, semiannually
fluvastatin 40 mg bid	-36%	B, 8 wk
fluvastatin XL 80 mg	-35%	B, 8 wk
lovastatin 40 mg	-31%	B, 6 & 12 wk, semiannually
pravastatin 40mg	-34%	B, prior to dose increase
rosuvastatin 5 mg	-45%	B, 12 wk, semiannually
simvastatin 20 mg	-38%	B, ***

*Adapted from *Circulation* 2004;110:227-239. Data taken from prescribing information for primary hypercholesterolemia. B=baseline, LDL=low-density lipoprotein, LFT=liver function tests. Will get ~6% decrease in LDL with every doubling of dose. **From prescribing info. ACC/AHA/NHLBI schedule for LFT monitoring: baseline, ~12 weeks after starting therapy, annually, when clinically indicated. Stop statin therapy if LFTs are >3 times upper limit of normal. ***Get LFTs prior to & 3 months after dose increase to 80 mg, then semiannually for first year.

Anti-Hyperlipidemic Agents - Other

bezafibrate (**✿***Bezalip*): Canada only. Hyperlipidemia/hypertriglyceridemia: 200 mg immediate release PO bid-tid, or 400 mg of sustained release PO daily. [Trade only: Immediate release tab: 200 mg. Sust'd release tab: 400 mg.] ▶K ♀D ▶- $$$

ezetimibe (*Zetia*, **✿***Ezetrol*): Hyperlipidemia: 10 mg PO daily. [Trade only: Tabs non-scored 10 mg] ▶L ♀C ▶? $$$

fenofibrate (*Tricor, Lipofen, Lofibra, Antara, Triglide*, **✿***Lipidil Micro, Lipidil Supra*): Hypertriglyceridemia: Tricor tabs: 48-145 mg PO daily, max 145 mg daily. Triglide: 50-160 mg PO daily, max 160 mg daily. Lipofen: 50-150 mg daily, max 150 mg daily. Lofibra/generic tabs: 54-160 mg, max 160 mg daily. Lofibra/generic micronized caps: 67-200 mg daily; max 200 mg daily. Antara micronized caps: 43-130 mg PO daily; max 130 mg daily. Hypercholesterolemia/mixed dyslipidemia: Tricor tabs: 145 mg PO daily. Triglide: 160 mg daily. Lipofen: 150 mg daily. Generic tabs: 160 mg daily. Lofibra or generic micronized caps 200 mg PO daily. Antara, micronized caps: 130 mg PO daily. Micronized caps should be taken with food. [Generic tabs, 54, 160 mg. Generic/Trade (Lofibra) micronized caps, 67, 134, 200 mg. Trade only: Tricor tabs 48,145 mg. Triglide tabs 50, 160 mg. Antara micronized caps 43, 130 mg. Lipofen tabs 50,100,150 mg.] ▶LK ♀C ▶- $$$

gemfibrozil (*Lopid*): Hypertriglyceridemia / primary prevention of coronary artery disease: 600 mg PO bid 30 minutes before meals. [Generic/Trade: Tabs, scored 600 mg.] ▶LK ♀C ▶? $$

niacin (**nicotinic acid, vitamin B3, Niacor, Nicolar, Niaspan**): Hyperlipidemia: Start 50-100 mg PO bid-tid with meals, increase slowly, usual maintenance range 1.5-3 g/day, max 6 g/day. Extended-release (Niaspan): Start 500 mg qhs, ↑ prn

LDL CHOLESTEROL GOALS[1]		Lifestyle Changes[2]	Also Consider Meds at LDL (mg/dL)[3]
Risk Category	LDL Goal		
High risk: CHD or equivalent risk,[4] 10-year risk >20%	<100 (optional < 70)*	LDL ≥100**	≥100 (<100: consider Rx options)#
Moderately-high risk: 2+ risk factors,[5] 10-year risk 10-20%	<130 mg/dL	LDL ≥130**	≥130 (100-129: consider Rx options)##
Moderate risk: 2+ risk factors,[5] 10-year risk <10%	<130 mg/dL	LDL ≥130	≥160
Lower risk: 0 to 1 risk factor[5]	<160 mg/dL	LDL ≥160	≥190 (160-189: Rx optional)

1. CHD=coronary heart disease. LDL=low density lipoprotein. Adapted from NCEP: *JAMA* 2001; 285:2486; NCEP Report: Circulation 2004;110:227-239. All 10-year risks based upon Framingham stratification; calculator available at: http://hin.nhlbi.nih.gov/atpiii/calculator.asp?usertype=prof. **2.** Dietary modification, weight reduction, exercise. **3.** When using LDL lowering therapy, achieve at least 30-40% LDL reduction. **4.** Equivalent risk defined as diabetes, other atherosclerotic disease (peripheral artery disease, abdominal aortic aneurysm, symptomatic carotid artery disease), or ≥2 risk factors such that 10 year risk >20%. **5.** Risk factors: Cigarette smoking, HTN (BP≥140/90 mmHg or on antihypertensive meds), low HDL (<40 mg/dL), family hx of CHD (1° relative: ♂ <55 yo, ♀ <65 yo), age (♂ ≥45 yo, ♀ ≥55 yo). *** For very high risk patients** (CVD with: acute coronary syndrome; diabetes; 2+ risk factors for metabolic syndrome; or severe/ poorly controlled risk factors), consider LDL goal <70. ******Regardless of LDL, lifestyle changes are indicated when lifestyle-related risk factors (obesity, physical inactivity, ↑TG, ↓ HDL, or metabolic syndrome) are present. **#** If baseline LDL <100, starting LDL lowering therapy is an option based on clinical trials. With ↑TG or ↓HDL, consider combining fibrate or nicotinic acid with LDL lowering drug. **##** At baseline or after lifestyle changes - initiating therapy to achieve LDL <100 is an option based on clinical trials.

monthly up to max 2000 mg. Extended-release formulations not listed here may have greater hepatotoxicity. Titrate slowly and use aspirin or NSAID 30 minutes before niacin doses to decrease flushing reaction. [Generic (OTC): Tabs, scored 25,50,100,150,500 mg. Trade only: Tabs, immediate-release, scored (Niacor), 500 mg; ext'd release, non-scored (Niaspan), 500,750,1000 mg.] ▶K ♀C ▶? $

Antihypertensive Combinations (See component drugs for ▶ ♀ ▶)

NOTE: Dosage should first be adjusted by using each drug separately.

BY TYPE: <u>ACE Inhibitor/Diuretic:</u> Accuretic, Capozide, Inhibace Plus, Lotensin HCT, Monopril HCT, Prinzide, Uniretic, Vaseretic, Zestoretic. <u>ACE Inhibitor/Calcium Channel Blocker:</u> Lexxel, Lotrel, Tarka. Angiotensin Receptor Blocker/Diuretic: Atacand HCT, Avalide, Benicar HCT, Diovan HCT, Hyzaar, Micardis HCT, Teveten HCT. <u>Beta-blocker/Diuretic:</u> Corzide, Inderide, Lopressor HCT, Tenoretic, Timolide, Ziac. <u>Diuretic combinations:</u> Aldactazide, Dyazide, Maxzide, Maxzide-25, Moduretic, Triazide. <u>Diuretic/miscellaneous antihypertensives:</u> Aldoril, Apresazide, Chlorpres, Minizide, Renese-R, Ser-Ap-Es. <u>Other:</u> BiDil.

BY NAME: **Accuretic** (quinapril + hydrochlorothiazide): Generic/Trade: Tabs, 10/12.5, 20/12.5, 20/25. **Aldactazide** (spironolactone + hydrochlorothiazide): Generic/Trade: Tabs, non-scored 25/25, scored 50/50 mg. **Aldoril** (methyldopa + hydrochlorothiazide): Generic/Trade: Tabs, non-scored, 250/15 (Aldoril-15), 250/25 mg (Aldoril-25). Tabs, non-scored, 500/30 (Aldoril D30), 500/50 mg (Aldoril D50). **Apresazide** (hydralazine + hydrochlorothiazide): Generic only: Caps 25/25, 50/50 mg. **Atacand HCT** (candesartan + hydrochlorothiazide, ♥*Atacand Plus*): Trade only: Tab, non-scored 16/12.5, 32/12.5 mg. **Avalide** (irbesartan + hydrochlorothiazide): Trade only: Tabs, non-scored 150/12.5, 300/12.5, 300/25 mg. **Benicar HCT** (olmesartan + hydrochlorothiazide): Trade only: Tabs, non-scored 20/12.5, 40/12.5, 40/25 mg. **BiDil** (hydralazine + isosorbide): Trade only: Tabs, scored 37.5/20 mg. **Capozide** (captopril + hydrochlorothiazide): Generic/Trade: Tabs, scored 25/15, 25/25, 50/15, 50/25 mg. **Clorpres** (clonidine + chlorthalidone): Trade only: Tabs, scored 0.1/15, 0.2/15, 0.3/15 mg. **Corzide** (nadolol + bendroflumethiazide): Trade only: Tabs, scored 40/5, 80/5 mg. **Diovan HCT** (valsartan + hydrochlorothiazide): Trade only: Tabs, non-scored 80/12.5, 160/12.5, 160/25, 320/12.5, 320/25 mg. **Dyazide** (triamterene + hydrochlorothiazide): Generic/Trade: Caps, (Dyazide) 37.5/25, (generic only) 50/25 mg. **Hyzaar** (losartan + hydrochlorothiazide): Trade only: Tabs, non-scored 50/12.5, 100/12.5, 100/25 mg. **Inderide** (propranolol + hydrochlorothiazide): Generic/Trade: Tabs, scored 40/25, 80/25. **Inhibace Plus** (cilazapril + hydrochlorothiazide): Trade only: Scored tabs 5 mg cilazapril + 12.5 mg HCTZ. **Lexxel** (enalapril + felodipine): Trade only: Tabs, non-scored 5/2.5,.5/5 mg. **Lopressor HCT** (metoprolol + hydrochlorothiazide): Generic/Trade: Tabs, scored 50/25, 100/25, 100/50 mg. **Lotensin HCT** (benazepril + hydrochlorothiazide): Generic/Trade: Tabs, scored 5/6.25, 10/12.5, 20/12.5, 20/25 mg. **Lotrel** (amlodipine + benazepril): Trade only: cap, 2.5/10, 5/10, 5 /20, 10/20, 5/40, 10/40 mg. **Maxzide** (triamterene + hydrochlorothiazide, ♥*Triazide*): Generic/Trade: Tabs, scored (Maxzide-25) 37.5/25 (Maxzide) 75/50 mg. **Maxzide-25** (triamterene + hydrochlorothiazide): Generic/Trade: Tabs, scored (Maxzide-25) 37.5/25 (Maxzide) 75/50 mg. **Micardis HCT** (telmisartan + hydrochlorothiazide, ♥*Micardis Plus*): Trade only: Tabs, non-scored 40/12.5, 80/12.5, 80/25 mg. **Minizide** (prazosin + polythiazide): Trade only: cap, 1/0.5, 2/0.5, 5/0.5 mg. **Moduretic** (amiloride + hydrochlorothiazide, ♥*Moduret*): Generic/Trade: Tabs, scored 5/50 mg. **Monopril HCT** (fosinopril + hydrochlorothiazide): Generic/Trade: Tabs, non-scored 10/12.5, scored

20/12.5 mg. **Prinzide** (*lisinopril + hydrochlorothiazide*): Generic/Trade: Tabs, non-scored 10/12.5, 20/12.5, 20/25 mg. **Renese-R** (*reserpine + polythiazide*): Trade only: Tabs, scored, 0.25/2 mg. **Ser-Ap-Es** (*hydralazine + hydrochlorothiazide + reserpine*): Generic: Tabs, non-scored, 25/15/0.1 mg. **Tarka** (*trandolapril + verapamil*): Trade only: Tabs, non-scored 2/180, 1/240, 2/240, 4/240 mg. **Tenoretic** (*atenolol + chlorthalidone*): Generic/Trade: Tabs, scored 50/25, non-scored 100/25 mg. **Teveten HCT** (*eprosartan + hydrochlorothiazide*): Trade only: Tabs, non-scored 600/12.5, 600/25 mg. **Timolide** (*timolol + hydrochlorothiazide*): Trade only: Tabs, non-scored 10/25 mg. **Uniretic** (*moexipril + hydrochlorothiazide*): Trade only: Tabs, scored 7.5/12.5, 15/12.5, 15/25 mg. **Vaseretic** (*enalapril + hydrochlorothiazide*): Generic/Trade: Tabs, non-scored 5/12.5, 10/25 mg. **Zestoretic** (*lisinopril + hydrochlorothiazide*): Generic/Trade: Tabs, non-scored 10/12.5, 20/12.5, 20/25 mg. **Ziac** (*bisoprolol + hydrochlorothiazide*): Generic/Trade: Tabs, non-scored 2.5/6.25, 5/6.25, 10/6.25 mg.

Antihypertensives - Other

epoprostenol (*Flolan*): Specialized dosing for pulmonary arterial HTN. ▶Plasma ♀B ▶? $$$$$

fenoldopam (*Corlopam*): Severe HTN: 10 mg in 250 mL D5W (40 mcg/mL), start at 0.1 mcg/kg/min titrate q15 min, usual effective dose 0.1-1.6 mcg/kg/min. ▶LK ♀B ▶? $$$$$

hydralazine (*Apresoline*): Hypertensive emergency: 10-50 mg IM or 10-20 mg IV, repeat as needed. HTN: Start 10 mg PO bid-qid, max 300 mg/day. Headaches, peripheral edema, lupus syndrome. [Generic/Trade: Tabs, non-scored 10,25,50, 100 mg.] ▶LK ♀C ▶+ $

nitroprusside (*Nipride, Nitropress*): Hypertensive emergency: 50 mg in 250 mL D5W (200 mcg/mL), start at 0.3 mcg/kg/min (for 70 kg adult = 6 mL/h). Max 10 mcg/kg/min. Protect from light. Cyanide toxicity with high doses, hepatic/renal impairment, and prolonged infusions; check thiocyanate levels. ▶RBC's ♀C ▶- $

phentolamine (*Regitine, Rogitine*): Diagnosis of pheochromocytoma: 5 mg increments IV/IM. Peds 0.05-0.1 mg/kg IV/IM up to 5 mg per dose. Extravasation: 5-10 mg in 10 mL NS local injection. ▶Plasma ♀C ▶? $$

Antiplatelet Drugs

abciximab (*ReoPro*): Platelet aggregation inhibition, percutaneous coronary intervention: 0.25 mg/kg IV bolus via separate infusion line before procedure, then 0.125 mcg/kg/min (max 10 mcg/min) IV infusion for 12h. ▶Plasma ♀C ▶? $$$$$

Aggrenox (*aspirin + dipyridamole*): Platelet aggregation inhibition: 1 cap bid. [Trade only: Caps, 25/200 mg aspirin/extended-release dipyridamole.] ▶LK ♀D ▶? $$$$

aspirin (*Ecotrin, Empirin, Halfprin, Bayer, ASA, ♥Entrophen, Asaphen, Novasen*): Platelet aggregation inhibition: 81-325 mg PO daily. [Generic/Trade (OTC): tabs, 325,500 mg; chewable 81 mg; enteric-coated 81,162 mg (Halfprin), 81,325,500 mg (Ecotrin), 650,975 mg. Trade only: tabs, controlled-release 650, 800 mg (ZORprin, Rx). Generic only (OTC): suppository 120,200,300,600 mg.] ▶K ♀D ▶? $

clopidogrel (*Plavix*): Reduction of thrombotic events: recent AMI/stroke, established peripheral arterial disease: 75 mg PO daily; acute coronary syndrome (unstable angina or non Q-wave MI): 300 mg loading dose, then 75 mg PO daily in combination with aspirin for 9-12 months. [Generic/Trade: Tab, non-scored 75 mg.] ▶LK ♀B ▶? $$$$

dipyridamole (***Persantine***): Antithrombotic: 75-100 mg PO qid. [Generic/Trade: Tabs, non-scored 25, 50, 75 mg.] ▶L ♀B ▶? $

eptifibatide (***Integrilin***): Acute coronary syndrome: Load 180 mcg/kg IV bolus, then infusion 2 mcg/kg/min for up to 72 hr. Discontinue infusion prior to CABG. Percutaneous coronary intervention: Load 180 mcg/kg IV bolus just before procedure, followed by infusion 2 mcg/kg/min and a second 180 mcg/kg IV bolus 10 min after the first bolus. Continue infusion for up to 18-24 hr (minimum 12 hr) after procedure. Reduce infusion dose with CrCl <50 mL/min; contraindicated in dialysis patients. ▶K ♀B ▶? $$$$$

ticlopidine (***Ticlid***): Due to high incidence of neutropenia and thrombotic thrombocytopenia purpura, other drugs preferred. Platelet aggregation inhibition/reduction of thrombotic stroke: 250 mg PO bid with food. [Generic/Trade: Tab, non-scored 250 mg.] ▶L ♀B ▶? $$$$

tirofiban (***Aggrastat***): Acute coronary syndromes: Start 0.4 mcg/kg/min IV infusion for 30 mins, then decrease to 0.1 mcg/kg/min for 48-108 hr or until 12-24 hr after coronary intervention. Half dose with CrCl <30 mL/min. Use concurrent heparin to keep PTT twice normal. ▶K ♀B ▶? $$$$$

Beta Blockers

> NOTE: See also antihypertensive combinations. Abrupt discontinuation may precipitate angina, myocardial infarction, arrhythmias, or rebound hypertension; discontinue by tapering over 2 weeks. Avoid using nonselective beta-blockers and use agents with beta1 selectivity cautiously in asthma/COPD. Beta 1 selectivity diminishes at high doses. Avoid in decompensated heart failure.

acebutolol (***Sectral***, ♣***Rhotral, Monitan***): HTN: Start 400 mg PO daily or 200 mg PO bid, maximum 1200 mg/day. Beta1 receptor selective. [Generic/Trade: Caps, 200,400 mg.] ▶LK ♀B ▶- $

atenolol (***Tenormin***): Acute MI: 5 mg IV over 5 min, repeat in 10 min. Start 25-50 mg PO daily or divided bid, maximum 100 mg/day. Beta1 receptor selective. [Generic/Trade: Tabs, non-scored 25,100; scored, 50 mg.] ▶K ♀D ▶- $

betaxolol (***Kerlone***): HTN: Start 5-10 mg PO daily, max 20 mg/day. Beta1 receptor selective. [Trade only: Tabs, scored 10, non-scored 20 mg.] ▶LK ♀C ▶? $$

bisoprolol (***Zebeta***, ♣***Monocor***): HTN: Start 2.5-5 mg PO daily, max 20 mg/day. Beta1 receptor selective. [Generic/Trade: Tabs, scored 5, non-scored 10 mg.] ▶LK ♀C ▶? $$

carvedilol (***Coreg***): Heart failure: Start 3.125 mg PO bid with food, double dose q2 weeks as tolerated up to max of 25 mg bid (if <85 kg) or 50 mg bid (if >85 kg). LV dysfunction following acute MI: Start 6.25 mg PO bid, double dose q 3-10 days as tolerated to max of 25 mg bid. HTN: Start 6.25 mg PO bid, double dose q7-14 days as tolerated to max 50 mg/day. Alpha1, beta1, and beta2 receptor blocker. [Trade only: Tabs, non-scored 3.125, 6.25, 12.5, 25 mg.] ▶LK ♀C ▶? $$$$

esmolol (***Brevibloc***): SVT/HTN emergency: Mix infusion 5 g in 500 mL (10 mg/mL), load with 500 mcg/kg over 1 minute (70 kg: 35 mg or 3.5 mL) then infusion 50-200 mcg/kg/min (70 kg: 100 mcg/kg/min = 40 mL/h). Half-life = 9 minutes. Beta1 receptor selective. ▶K ♀C ▶? $$$

labetalol (***Trandate***): HTN: Start 100 mg PO bid, max 2400 mg/day. HTN emergency: Start 20 mg IV slow injection, then 40-80 mg IV q10 min prn up to 300 mg or IV infusion 0.5-2 mg/min. Peds: Start 0.3-1 mg/kg/dose (max 20 mg). Alpha1, beta1, and beta2 receptor blocker. [Generic/Trade: Tabs, scored 100,200,300 mg.] ▶LK ♀C ▶+ $$$

metoprolol (***Lopressor, Toprol-XL***, ♣***Betaloc***): Acute MI: 5 mg increments IV q5-15 min up to 15 mg followed by oral therapy. HTN (immediate release): Start 100 mg PO daily or in divided doses, increase as needed up to 450 mg/day; may require multiple daily doses to maintain 24 hour BP control. HTN (extended-release): Start 25-100 mg PO daily, increase as needed up to 400 mg/day. Heart failure: Start 12.5-25 mg (extended-release) PO daily, double dose every 2 weeks as tolerated up to max 200 mg/day. Angina: Start 50 mg PO bid (immediate-release) or 100 mg PO daily (extended-release), increase as needed up to 400 mg/day. Beta1 receptor selective. [Generic/Trade: tabs, scored 50,100 mg. Trade only: tabs, extended-release, scored (Toprol-XL) 25, 50, 100, 200 mg. Generic only: tabs, scored 25 mg.] ▶L ♀C ▶? $$

nadolol (**Corgard**): HTN: Start 20-40 mg PO daily, max 320 mg/day. Beta1 and beta2 receptor blocker. [Generic/Trade: Tabs, scored 20,40,80,120,160 mg.] ▶K ♀C ▶- $$

oxprenolol (✤**Trasicor, Slow-Trasicor**): Canada only. HTN: Regular release: Initially 20 mg PO tid, titrate upwards prn to usual maintenance 120-320 mg/day divided bid-tid. Alternatively, may substitute an equivalent daily dose of sustained release product; do not exceed 480 mg/day. [Trade only: Regular release tabs: 40, 80 mg. Sustained release tabs: 80, 160 mg.] ▶L ♀D ▶- $$

penbutolol (**Levatol**): HTN: Start 20 mg PO daily, max 80 mg/day. [Trade only: Tabs, scored 20 mg.] ▶LK ♀C ▶? $$$

pindolol (✤**Visken**): HTN: Start 5 mg PO bid, max 60 mg/day. Beta1 and beta2 receptor blocker. [Generic: Tabs, scored 5,10 mg.] ▶K ♀B ▶? $$$

propranolol (**Inderal, Inderal LA, InnoPran XL**): HTN: Start 20-40 mg PO bid or 60-80 mg PO daily; extended-release (Inderal LA) max 640 mg/day; extended release (InnoPran XL) 80 mg qhs (10 PM), max 120 mg qhs. Supraventricular tachycardia or rapid atrial fibrillation/flutter: 1 mg IV q2min. Max of 2 doses in 4 hours. Migraine prophylaxis: Start 40 mg PO bid or 80 mg PO daily (extended-release), max 240 mg/day. Beta1 and beta2 receptor blocker. [Generic/Trade: Tabs, scored 10, 20, 40, 60, 80. Generic only: Solution 20, 40 mg/5 mL. Concentrate 80 mg/mL. Trade only: Caps, extended-release (Inderal LA daily) 60, 80, 120, 160 mg; (InnoPran XL qhs) 80, 120 mg.] ▶L ♀C ▶+ $$

timolol (**Blocadren**): HTN: Start 10 mg PO bid, max 60 mg/d. Beta1 and beta2 blocker. [Generic/Trade: Tabs, non-scored 5, scored 10,20 mg.] ▶LK ♀C ▶+ $$

Calcium Channel Blockers (CCBs) – Dihydropyridines (See also combinations)

amlodipine (**Norvasc**): HTN: Start 2.5 to 5 mg PO daily, max 10 daily. [Generic/Trade: tabs, non-scored 2.5,5,10 mg.] ▶L ♀C ▶? $$$

felodipine (**Plendil, ✤Renedil**): HTN: Start 2.5-5 mg PO daily, max 10 mg/day. [Generic/Trade: Tabs, extended-release, non-scored 2.5,5,10 mg.] ▶L ♀C ▶? $$$

isradipine (**DynaCirc, DynaCirc CR**): HTN: Start 2.5 mg PO bid, max 20 mg/day (max 10 mg/day in elderly). Controlled-release: 5-10 mg PO daily. [Generic/Trade: Immediate release caps 2.5 mg. Trade only: Tabs, controlled-release 5,10 mg.] ▶L ♀C ▶? $$$

nicardipine (**Cardene, Cardene SR**): HTN emergency: Begin IV infusion at 5 mg/h, titrate to effect, max 15 mg/h. HTN: Start 20 mg PO tid, max 120 mg/day. Sustained release: Start 30 mg PO bid, max 120 mg/day. [Generic/Trade: caps 20,30 mg. Trade only: caps, sustained-release 30,45,60 mg.] ▶L ♀C ▶? $$$

nifedipine (**Procardia, Adalat, Procardia XL, Adalat CC, ✤Adalat XL, Adalat PA**): HTN/angina: extended-release: 30-60 mg PO daily, max 120/d. Angina: immediate-release: Start 10 mg PO tid, max 120 mg/d. Avoid sublingual administration, may cause excessive hypotension, AMI, stroke. Do not use immediate-release caps for treating HTN. [Generic/Trade: Caps, 10,20 mg. Tabs, extended-release 30,60,90 mg.] ▶L ♀C ▶+ $$

nisoldipine (*Sular*): HTN: Start 20 mg PO daily, max 60 mg/day. [Trade only: Tabs, extended-release 10,20,30,40 mg.] ▶L ♀C ▶? $$

Calcium Channel Blockers (CCBs) – Other (See also antihypertensive combinations)

diltiazem (*Cardizem, Cardizem SR, Cardizem LA, Cardizem CD, Cartia XT, Dilacor XR, Diltiazem CD, Diltia XT, Diltzac, Tiazac, Taztia XT*): Atrial fibrillation/flutter, PSVT: bolus 20 mg (0.25 mg/kg) IV over 2 min. Rebolus 15 min later (if needed) 25 mg (0.35 mg/kg). Infusion 5-15 mg/h. Once daily, extended-release, HTN: Start 120-240 mg PO daily, max 540 mg/day. Once daily, graded extended-release (Cardizem LA), HTN: Start 180-240 mg PO daily, max 540 mg/day. Twice daily, sustained-release, HTN: Start 60-120 mg PO bid, max 360 mg/day. Immediate-release, angina: Start 30 mg PO qid, max 360 mg/day divided tid-qid; extended-release, Start 120-240 mg PO daily, max 540 mg/day. Once daily, graded extended-release (Cardizem LA), angina: Start 180 mg PO daily, doses >360 mg may provide no additional benefit. [Generic/Trade: Immediate-release tabs, non-scored (Cardizem) 30, scored 60, 90, 120 mg; extended-release caps (Cardizem CD, Taztia XT daily) 120, 180, 240, 300, 360 mg, (Cartia XT, Dilt-CD daily) 120, 180, 240, 300, (Tiazac daily) 120, 180, 240, 300, 360, 420 mg, (Dilacor XR, Diltia XT daily) 120, 180, 240 mg, (Diltzac) 120,180,240,300,360 mg. Trade only: Sustained-release caps (Cardizem SR q12h) 60, 90, 120 mg; extended-release graded tabs (Cardizem LA daily) 120, 180, 240, 300, 360, 420 mg.] ▶L ♀C ▶+ $$

verapamil (*Isoptin, Calan, Covera-HS, Verelan, Verelan PM, ✦Chronovera, Veramil*): SVT: 5-10 mg IV over 2 min; peds (1-15 yo): 2-5 mg (0.1-0.3 mg/kg) IV, max dose 5 mg. Angina: immediate-release, start 40-80 mg PO tid-qid, max 480 mg/day; sustained-release, start 120-240 mg PO daily, max 480 mg/day (use bid dosing for doses >240 mg/day with Isoptin SR and Calan SR); (Covera-HS) 180 mg PO qhs, max 480 mg/day. HTN: same as angina, except (Verelan PM) 100-200 mg PO qhs, max 400 mg/day; immediate-release tabs should be avoided in treating HTN. [Generic/Trade: tabs, immediate-release, scored 40,80,120 mg; sustained-release, non-scored (Calan SR, Isoptin SR) 120, scored 180,240 mg; caps, sustained-release (Verelan) 120,180,240,360 mg. Trade only: tabs, extended-release (Covera HS) 180,240 mg; caps, extended-release (Verelan PM) 100,200,300 mg.] ▶L ♀C ▶+ $$

Diuretics - Carbonic Anhydrase Inhibitors

acetazolamide (*Diamox*): Acute mountain sickness: 125-250 mg PO bid-tid, beginning 1-2 days prior to ascent and continuing for ≥5 days at higher altitude. [Generic only: Tabs, 125,250 mg; Trade: cap, sust'd-release 500 mg.] ▶LK ♀C ▶+ $

Diuretics - Loop

bumetanide (*Bumex, ✦Burinex*): Edema: 0.5-1 mg IV/IM; 0.5-2 mg PO daily. 1 mg bumetanide is roughly equivalent to 40 mg furosemide. [Generic/Trade: Tabs, scored 0.5,1,2 mg.] ▶K ♀C ▶? $

ethacrynic acid (*Edecrin*): Rarely used. May be useful in sulfonamide-allergic patients. Edema: 0.5-1.0 mg/kg IV, max 50 mg; 25-100 mg PO daily-bid. [Trade only: Tabs, scored 25] ▶K ♀B ▶? $

furosemide (*Lasix*): Edema: Initial dose 20-80 mg IV/IM/PO, increase dose by 20-

40 mg every 6-8h until desired response is achieved, max 600 mg/day. Use lower doses in elderly. [Generic/Trade: Tabs, non-scored 20, scored 40,80 mg. Generic only: Oral solution 10 mg/mL, 40 mg/5 mL.] ▶K ♀C ▶? $

torsemide (*Demadex*): Edema: 5-20 mg IV/PO daily. [Generic/Trade: Tabs, scored 5,10,20,100 mg.] ▶LK ♀B ▶? $

Diuretics - Potassium Sparing

amiloride (*Midamor*): Diuretic-induced hypokalemia: Start 5 mg PO daily, max 20 mg/day. [Generic/Trade: Tabs, non-scored 5 mg.] ▶LK ♀B ▶? $$

Diuretics - Thiazide Type (See also antihypertensive combinations.)

chlorothiazide (*Diuril*): HTN: 125-250 mg PO daily or divided bid, max 1000 mg/day divided bid. Edema: 500-2000 mg PO/IV daily or divided bid. [Generic: Tabs, scored 250,500 mg.] ▶L ♀C, D if used in pregnancy-induced HTN ▶+ $

chlorthalidone (*Thalitone*): HTN: 12.5-25 mg PO daily, max 50 mg/day. Edema: 50-100 mg PO daily, max 200 mg/day. Nephrolithiasis (unapproved use): 25-50 mg PO daily. [Trade only: Tabs, non-scored (Thalitone) 15 mg. Generic only: Tabs non-scored 25,50 mg.] ▶L ♀B, D if used in pregnancy-induced HTN ▶+ $

hydrochlorothiazide (*HCTZ, Esidrix, Oretic, Microzide*): HTN: 12.5-25 mg PO daily, max 50 mg/day. Edema: 25-100 mg PO daily, max 200 mg/day. [Generic/Trade: Tabs, scored 25, 50 mg; Cap 12.5 mg.] ▶L ♀B, D if used in pregnancy-induced HTN ▶+ $

indapamide (*Lozol*, ♣*Lozide*): HTN: 1.25-5 mg PO daily, max 5 mg/day. Edema: 2.5-5 mg PO qam. [Generic/Trade: Tabs, non-scored 1.25, 2.5 mg.] ▶L ♀B, D if used in pregnancy-induced HTN ▶? $

metolazone (*Zaroxolyn*): Edema: 5-10 mg PO daily, max 10 mg/day in heart failure, 20 mg/day in renal disease. If used with loop diuretic, start with 2.5 mg PO daily. [Generic/Trade: tabs 2.5,5,10 mg.] ▶L ♀B, D if used in pregnancy-induced HTN ▶? $$$

Nitrates

isosorbide dinitrate (*Isordil, Dilatrate-SR*, ♣*Cedocard SR, Coronex*): Angina prophylaxis: 5-40 mg PO tid (7 am, noon, 5 pm), sustained-release: 40-80 mg PO bid (8 am, 2 pm). Acute angina, SL Tabs: 2.5-10 mg SL q5-10 min prn, up to 3 doses in 30 min. [Generic/Trade: Tabs, scored 5,10,20,30,40, chewable, scored 5,10, sublingual tabs, scored 2.5,5,10 mg. Trade only: cap, sustained-release (Dilatrate-SR) 40 mg. Generic only: tab, sustained-release 40 mg.] ▶L ♀C ▶? $

isosorbide mononitrate (*ISMO, Monoket, Imdur*): Angina: 20 mg PO bid (8 am and 3 pm). Extended-release: Start 30-60 mg PO daily, maximum 240 mg/day. [Generic/Trade: Tabs, non-scored (ISMO, bid dosing) 20, scored (Monoket, bid dosing) 10,20, extended-release, scored (Imdur, daily dosing) 30,60, non-scored 120 mg.] ▶L ♀C ▶? $$

nitroglycerin intravenous infusion (*Tridil*): Perioperative HTN, acute MI/Heart failure, acute angina: mix 50 mg in 250 mL D5W (200 mcg/mL), start at 10-20 mcg/min (3-6 mL/h), then titrate upward by 10-20 mcg/min as needed. [Brand name "Tridil" no longer manufactured, but retained herein for name recognition.] ▶L ♀C ▶? $

nitroglycerin ointment (*Nitrol, Nitro-BID*): Angina prophylaxis: Start 0.5 inch q8h, maintenance 1-2 inches q8h, max 4 inches q4-6h; 15 mg/inch. Allow for a nitrate-free period of 10-14 h to avoid nitrate tolerance. 1 inch ointment is approximately 15 mg. [Trade only: Ointment, 2%, tubes 1,30,60g (Nitro-BID).] ▶L ♀C ▶? $

nitroglycerin spray (***Nitrolingual***): Acute angina: 1-2 sprays under the tongue prn, max 3 sprays in 15 min. [Trade only: Solution, 4,9,12 mL. 0.4 mg/spray (60 or 200 sprays/canister).] ▶L ♀C ▶? $$$

nitroglycerin sublingual (***Nitrostat, NitroQuick***): Acute angina: 0.4 mg SL under tongue, repeat dose every 5 min as needed up to 3 doses in 15 min. [Generic/Trade: Sublingual tabs, non-scored 0.3,0.4,0.6 mg; in bottles of 100 or package of 4 bottles with 25 tabs each.] ▶L ♀C ▶? $

nitroglycerin sustained release: Angina prophylaxis: Start 2.5 mg PO bid-tid, then titrate upward prn. Allow for a nitrate-free period of 10-14 h to avoid nitrate tolerance. [Generic only: cap, extended-release 2.5, 6.5, 9 mg.] ▶L ♀C ▶? $

nitroglycerin transdermal (***Minitran, Nitro-Dur, ✦Trinipatch***): Angina prophylaxis: 1 patch 12-14 h each day. Allow for a nitrate-free period of 10-14 h each day to avoid nitrate tolerance. [Trade only: Transdermal system, doses in mg/h: Nitro-Dur 0.1,0.2,0.3,0.4,0.6,0.8; Minitran 0.1,0.2,0.4,0.6. Generic only: Transdermal system, doses in mg/h: 0.1,0.2,0.4,0.6.] ▶L ♀C ▶? $$$

nitroglycerin transmucosal (***Nitroguard***): Acute angina and prophylaxis: 1-3 mg PO, between lip and gum or between cheek and gum, q 3-5 hours while awake. [Trade only: Tabs, controlled-release 2,3 mg.] ▶L ♀C ▶? $$$

Pressors / Inotropes

dobutamine (***Dobutrex***): Inotropic support: 2-20 mcg/kg/min. 70 kg: 5 mcg/kg/min with 1 mg/mL concentration (eg, 250 mg in 250 mL D5W) = 21 mL/h. ▶Plasma ♀D ▶- $

dopamine (***Intropin***): Pressor: Start at 5 mcg/kg/min, increase as needed by 5-10 mcg/kg/min increments at 10 min intervals, max 50 mcg/kg/min. 70 kg: 5 mcg/kg/min with 1600 mcg/mL concentration (eg, 400 mg in 250 mL D5W) = 13 mL/h. Doses in mcg/kg/min: 2-4 = (traditional renal dose, apparently ineffective) dopaminergic receptors; 5-10 = (cardiac dose) dopaminergic and beta1 receptors; >10 = dopaminergic, beta1, and alpha1 receptors. ▶Plasma ♀C ▶- $

ephedrine: Pressor: 10-25 mg slow IV, repeat q5-10 min prn. [Generic: Caps, 25,50 mg.] ▶K ♀C ▶? $

epinephrine (***EpiPen, EpiPen Jr, Twinject,* adrenalin**): Cardiac arrest: 1 mg IV q3-5 minutes. Anaphylaxis: 0.1-0.5 mg SC/IM, may repeat SC dose q 10-15 minutes. Acute asthma & hypersensitivity reactions: Adults: 0.1 to 0.3 mg of 1:1,000 soln SC or IM; Peds: 0.01 mg/kg (up to 0.3 mg) of 1:1,000 soln SC or IM. [Soln for injection: 1,10,000 (1 mg/mL in 10 mL syringe), 1,1,000 (1 mg/mL in 1 mL amps). EpiPen Auto-injector delivers 0.3 mg (1:1,000 soln) IM dose. EpiPen Jr. Autoinjector delivers 0.15 mg (1:2,000 solution) IM dose. Twinject Auto-injector delivers 0.15, 0.3 mg IM/SQ dose.] ▶Plasma ♀C ▶- $

CARDIAC PARAMETERS AND FORMULAS	Normal
Cardiac output (CO) = heart rate x stroke volume	4-8 l/min
Cardiac index (CI) = CO/BSA	2.8-4.2 l/min/m2
MAP (mean arterial press) = [(SBP - DBP)/3] + DBP	80-100 mmHg
SVR (systemic vasc resis) = (MAP - CVP)(80)/CO	800-1200 dyne/sec/cm5
PVR (pulm vasc resis) = (PAM - PCWP)(80)/CO	45-120 dyne/sec/cm5
QT_C = QT / square root of RR [calculate using both measures in sec]	≤0.44
Right atrial pressure (central venous pressure)	0-8 mmHg
Pulmonary artery systolic pressure (PAS)	20-30 mmHg
Pulmonary artery diastolic pressure (PAD)	10-15 mmHg
Pulmonary capillary wedge pressure (PCWP)	8-12 mmHg (post-MI ~16 mmHg)

inamrinone (**Amrinone**): Heart failure: 0.75 mg/kg bolus IV over 2-3 min, then infusion 100 mg in 100 mL NS (1 mg/mL) at 5-10 mcg/kg/min. 70 kg: 5 mcg/kg/min = 21 mL/h. ▶K ♀C ▶? $$$$$

midodrine (**Orvaten, ProAmatine**, ✚**Amatine**): Orthostatic hypotension: 10 mg PO tid while awake. [Generic/Trade: Tabs, scored 2.5,5,10 mg.] ▶LK ♀C ▶? $$$$$

milrinone (**Primacor**): Systolic heart failure (NYHA class III,IV): Load 50 mcg/kg IV over 10 min, then begin IV infusion of 0.375-0.75 mcg/kg/min. ▶K ♀C ▶? $$$$$

norepinephrine (**Levophed**): Acute hypotension: 4 mg in 500 mL D5W (8 mcg/mL) start 8-12 mcg/min, adjust to maintain BP, average maintenance rate 2-4 mcg/min, ideally through central line. 3 mcg/min = 22.5 mL/h. ▶Plasma ♀C ▶? $

phenylephrine (**Neo-Synephrine**): Severe hypotension: 50 mcg boluses IV. Infusion: 20 mg in 250 mL D5W (80 mcg/mL), start 100-180 mcg/min (75-135 mL/h), usual dose once BP is stabilized 40-60 mcg/min. ▶Plasma ♀C ▶- $

Thrombolytics

alteplase (**tpa, t-PA, Activase, Cathflo**, ✚**Activase rt-PA**): Acute MI: 15 mg IV bolus, then 50 mg over 30 min, then 35 mg over the next 60 min; (patient ≤67 kg) 15 mg IV bolus, then 0.75 mg/kg (max 50 mg) over 30 min, then 0.5 mg/kg (max 35 mg) over the next 60 min. Concurrent heparin infusion. Acute ischemic stroke: 0.9 mg/kg up to 90 mg infused over 60 min, with 10% of dose as initial IV bolus over 1 minute; start within 3 h of symptom onset. Acute pulmonary embolism: 100 mg IV over 2h, then restart heparin when PTT ≤twice normal. Occluded central venous access device: 2 mg/mL in catheter for 2 hr. May use second dose if needed. ▶L ♀C ▶? $$$$$

reteplase (**Retavase**): Acute MI: 10 units IV over 2 min; repeat in 30 min. ▶L ♀C ▶? $$$$$

streptokinase (**Streptase, Kabikinase**): Acute MI: 1.5 million units IV over 60 minutes. ▶L ♀C ▶? $$$$$

tenecteplase (**TNKase**): Acute MI: Single IV bolus dose over 5 seconds based on body weight; <60 kg, 30 mg; 60-69 kg, 35 mg; 70-79 kg, 40 mg; 80-89 kg, 45 mg; ≥90kg, 50 mg. ▶L ♀C ▶? $$$$$

urokinase (**Abbokinase, Abbokinase Open -Cath**): PE: 4400 units/kg loading dose over 10 min, followed by IV infusion 4400 units/kg/h for 12 hours. Occluded IV catheter: 5000 units instilled into catheter, remove solution after 5 min. ▶L ♀B ▶? $$$$$

THROMBOLYTIC THERAPY FOR ACUTE MI (if high-volume cath lab unavailable) *Indications*: Clinical history & presentation strongly suggestive of MI within 12 hours plus ≥1 of the following: 1 mm ST↑ in ≥2 contiguous leads; new left BBB; or 2 mm ST↓ in V1-4 suggestive of true posterior MI. *Absolute contraindications*: Previous cerebral hemorrhage, known cerebral aneurysm or arteriovenous malformation, known intracranial neoplasm, recent (<3 months) ischemic stroke (except acute ischemic stroke <3 hours), aortic dissection, active bleeding or bleeding diathesis (excluding menstruation), significant closed head or facial trauma (<3 months). *Relative contraindications*: Severe uncontrolled HTN (>180/110 mmHg) on presentation or chronic severe HTN; prior ischemic stroke (>3 months), dementia, other intracranial pathology; traumatic/prolonged (>10 minutes) cardiopulmonary resuscitation; major surgery (<3 weeks); recent (within 2-4 weeks) internal bleeding; puncture of non-compressible vessel; pregnancy; active peptic ulcer disease; current use of anticoagulants. For streptokinase/anistreplase: prior exposure (>5 days ago) or prior allergic reaction.   *Circulation* 2004;110:588-636

Volume Expanders

albumin (Albuminar, Buminate, Albumarc, ♣Plasbumin): Shock, burns: 500 mL of 5% solution IV infusion as rapidly as tolerated, repeat in 30 min if needed. ▶L ♀C ▶? $$$$

dextran (Rheomacrodex, Gentran, Macrodex): Shock/hypovolemia: 20 mL/kg up to 500 mL IV. ▶K ♀C ▶? $$$$

hetastarch (Hespan, Hextend): Shock/hypovolemia: 500-1000 mL IV 6% solution. ▶K ♀C ▶? $$$

plasma protein fraction (Plasmanate, Protenate, Plasmatein): Shock/hypovolemia: 5% soln 250-500 mL IV prn. ▶L ♀C ▶? $$$

Other

cilostazol (Pletal): Intermittent claudication: 100 mg PO bid on empty stomach. 50 mg PO bid with cytochrome P450 3A4 inhibitors (eg, ketoconazole, itraconazole, erythromycin, diltiazem) or cytochrome P450 2C19 inhibitors (eg, omeprazole). Avoid grapefruit juice. [Generic/Trade: Tabs 50, 100 mg.] ▶L ♀C ▶? $$$

nesiritide (Natrecor): Hospitalized patients with decompensated heart failure with dyspnea at rest: 2 mcg/kg IV bolus over 60 seconds, then 0.01 mcg/kg/min IV infusion for up to 48 hours. Do not initiate at higher doses. Limited experience with increased doses. 1.5 mg vial in 250mL D5W (6 mcg/mL). 70 kg: 2 mcg/kg bolus = 23.3 mL, 0.01 mcg/kg/min infusion = 7 mL/h. Symptomatic hypotension. May increase mortality. Not indicated for outpatient infusion, for scheduled repetitive use, to improve renal function, or to enhance diuresis. ▶K, plasma ♀C ▶? $$$$$

pentoxifylline (Trental): 400 mg PO tid with meals. [Generic/Trade: Tabs 400 mg.] ▶L ♀C ▶? $$$

ranolazine (Ranexa): Chronic angina: 500 mg PO bid, max 2000 mg daily. Reserve for angina not controlled with other antianginal drugs. Get baseline and follow-up EKGs; evaluate effects on QT interval. Contraindicated with pre-existing QT prolongation, hepatic impairment, QT prolonging drugs. Many drug interactions. [Trade: Tabs, extended release 500 mg.] ▶LK ♀C ▶? $$$$

CONTRAST MEDIA

MRI Contrast (all non-iodinated)

ferumoxides (Feridex): Iron-based IV contrast for hepatic MRI. ▶L ♀C ▶? $$$$

ferumoxsil (GastroMARK): Iron-based, oral GI contrast for MRI. ▶L ♀B ▶? $$$$

gadodiamide (Omniscan): IV contrast for MRI. ▶K ♀C ▶? $$$$

gadopentetate (Magnevist): IV contrast for MRI. ▶K ♀C ▶? $$$

gadoteridol (Prohance): IV contrast for MRI. ▶K ♀C ▶? $$$$

gadoversetamide (OptiMARK): IV contrast for MRI. ▶K ♀C ▶- $$$$

mangafodipir (Teslascan): IV contrast for MRI. ▶L ♀- ▶- $$$$

Radiography Contrast

> **NOTE:** Beware of allergic or anaphylactoid reactions. Avoid IV contrast in renal insufficiency or dehydration. Hold metformin (Glucophage) prior to or at the time of iodinated contrast dye use and for 48 h after procedure. Restart after procedure only if renal function is normal.

barium sulfate: Non-iodinated GI (eg, oral, rectal) contrast. ▶Not absorbed ♀? ▶+ $

diatrizoate (Cystografin, Gastrografin, Hypaque, MD-Gastroview, RenoCal, Reno-DIP, Reno-60, Renografin): Iodinated, ionic, high osmolality IV or GI contrast. ▶K ♀C ▶? $

iodixanol (Visipaque): Iodinated, non-ionic, low osm IV contrast. ▶K ♀B ▶? $$$

iohexol (*Omnipaque*): Iodinated, non-ionic, low osmolality IV and oral/body cavity contrast. ▸K ♀B ▸? $$$

iopamidol (*Isovue*): Iodinated, non-ionic, low osmolality IV contrast. ▸K ♀? ▸? $$

iopromide (*Ultravist*): Iodinated, non-ionic, low osm IV contrast. ▸K ♀B ▸? $$$

iothalamate (*Conray*, ♥*Vascoray*): Iodinated, ionic, high osmolality IV contrast. ▸K♀B ▸- $

ioversol (*Optiray*): Iodinated, non-ionic, low osmolality IV contrast. ▸K ♀B ▸? $$$

ioxaglate (*Hexabrix*): Iodinated, ionic, low osmolality IV contrast. ▸K ♀B ▸- $$$

ioxilan (*Oxilan*): Iodinated, non-ionic, low osmolality IV contrast. ▸K ♀B ▸- $$$

DERMATOLOGY

Acne Preparations

adapalene (*Differin*): apply qhs. [Trade only: gel 0.1% (15,45 g) cream 0.1% (15,45 g), soln 0.1% pad (30 mL).] ▸Bile ♀C ▸? $$

azelaic acid (*Azelex, Finacea, Finevin*): apply bid. [Trade only: cream 20%, 30g, 50g (Azelex, Finevin), gel 15% 30g (Finacea).] ▸K ♀B ▸? $$$

BenzaClin (clindamycin + benzoyl peroxide): apply bid. [Trade only: gel clindamycin 1% + benzoyl peroxide 5%; 25, 50 g.] ▸K ♀C ▸+ $$$

Benzamycin (erythromycin + benzoyl peroxide): apply bid. [Generic/Trade: gel erythromycin 3% + benzoyl peroxide 5%; 23.3, 46.6 g.] ▸LK ♀C ▸? $$$

benzoyl peroxide (*Benzac, Benzagel 10%, Desquam, Clearasil, ♥Acetoxyl, Solugel, Benoxyl, Oxyderm*): apply daily; increase to bid-tid if needed. [OTC and Rx generic: liquid 2.5,5,10%, bar 5,10%, mask 5%, lotion 5,5.5,10%, cream 5,10%, cleanser 10%, gel 2.5,4,5, 6,10,20%.] ▸LK ♀C ▸? $

Clenia (sodium sulfacetamide + sulfur): apply 1-3 times daily. [Generic: lotion (sodium sulfacetamide 10% & sulfur 5%) 25 g. Trade only (sodium sulfacetamide 10% & sulfur 5%): cream 28 g, foaming wash 170,340 g.] ▸K ♀C ▸? $$

clindamycin - topical (*Cleocin T, ClindaMax, Evoclin, ClindaMax, ♥Dalacin T*): apply daily (Evoclin) or bid (Cleocin T). [Generic/Trade: gel 1% 7.5, 30g, lotion 1% 60 mL, solution 1% 30,60 mL. Trade only: foam 1% 50 g (Evoclin).] ▸L ♀B ▸- $$$

Diane-35 (cyproterone + ethinyl estradiol): Canada only. 1 tab PO daily for 21 consecutive days, stop for 7 days, repeat cycle. [Rx trade only: blister pack of 21 tabs 2 mg/0.035 mg cyproterone acetate/ethinyl estradiol.] ▸L ♀X ▸- $$

Duac (clindamycin + benzoyl peroxide, ♥*Clindoxyl*): apply qhs. [Trade only: gel clindamycin 1% + benzoyl peroxide 5%; 45 g.] ▸K ♀C ▸+ $$$$

erythromycin - topical (*Eryderm, Erycette, Erygel, A/T/S, ♥Sans-Acne, Erysol*): apply bid. [Generic: solution 1.5% 60 mL, 2% 60,120 mL, pads 2%, gel 2% 30,60 g, ointment 2% 25 g.] ▸L ♀B ▸? $$

isotretinoin (*Accutane, Amnesteem, Claravis, Sotret, ♥Isotrex*): 0.5-2 mg/kg/day PO divided bid for 15-20 weeks. Typical target dose is 1 mg/kg/day. Can only be prescribed by healthcare professionals who are registered with the iPledge program. Potent teratogen; use extreme caution. May cause depression. Not for long term use. [Generic/Trade: caps 10, 20, 40 mg. Generic only (Sotret & Claravis): caps 30 mg.] ▸LK ♀X ▸- $$$$$

Rosula (sodium sulfacetamide + sulfur): apply 1-3 times daily. [Generic: lotion (sodium sulfacetamide 10% & sulfur 5%) 25 g. Trade only: gel (sodium sulfacetamide 10% & sulfur 5%) 45 mL, aqueous cleanser (sodium sulfacetamide 10% & sulfur 5%) 355 mL.] ▸K ♀C ▸? $$

sodium sulfacetamide - topical (*Klaron*): apply bid. [Trade only: lotion 10% 59, 118 mL.] ▸K ♀C ▸? $$

Sulfacet-R (sodium sulfacetamide + sulfur): apply 1-3 times daily. [Trade (Sulfacet-R) and generic: lotion sodium sulfacetamide 10% & sulfur 5%, 25 g] ▸K ♀C ▸? $$

tazarotene (*Tazorac, Avage*): Acne (Tazorac): apply 0.1% cream qhs. Psoriasis: apply 0.05% cream qhs, increase to 0.1% prn. [Trade only (Tazorac): Cream 0.05% and 0.1% - 30, 60g, Gel 0.05% and 0.1% - 30, 100g. Trade only (Avage): Cream 0.1% 15, 30g.] ▸L ♀X ▸? $$$

tretinoin (*Retin-A, Retin-A Micro, Renova, Retisol-A, ✿Stieva-A, Rejuva-A, Vitamin A Acid Cream*): Apply qhs. [Generic/Trade: cream 0.025% 20,45 g, 0.05% 20,45 g, 0.1% 20,45 g, gel 0.025% 15,45 g, 0.1% 15,45 g, liquid 0.05% 28 mL. Trade only: Renova cream 0.02% & 0.05% 40,60 g, Retin-A Micro gel 0.04%, 0.1% 20,45 g.] ▸LK ♀C ▸? $$$

Actinic Keratosis Preparations

diclofenac - topical (*Solaraze*): Actinic/solar keratoses: apply bid to lesions x 60-90 days. [Trade only: gel 3% 25, 50 g.] ▸L ♀B ▸? $$$$

fluorouracil - topical (*5-FU, Carac, Efudex, Fluoroplex*): Actinic keratoses: apply bid x 2-6 wks. Superficial basal cell carcinomas: apply 5% cream/solution bid. [Trade only: Cream 0.5% 30 g (Carac), 5% 25, 40 g (Efudex), 1% 30 g (Fluoroplex). Generic/Trade: Solution 2% & 5% 10 mL (Efudex).] ▸L ♀X ▸- $$$$$

Antibacterials

bacitracin (*✿Baciguent*): apply daily-tid. [OTC Generic/Trade: ointment 500 units/g 1,15,30g.] ▸Not absorbed ♀C ▸? $

fusidic acid (*Fucidin*): Canada only. Apply tid-qid. [Trade only: cream 2% fusidic acid 15,30 g, ointment 2% sodium fusidate 15,30 g.] ▸L ♀? ▸? $

gentamicin - topical (*Garamycin*): apply tid-qid. [Generic/Trade: ointment 0.1% 15,30 g, cream 0.1% 15,30 g.] ▸K ♀D ▸? $

mafenide (*Sulfamylon*): Apply daily-bid. [Trade only: cream 37, 114, 411 g, 5% topical solution 50 g packets.] ▸LK ♀C ▸? $$$

metronidazole - topical (*Noritate, MetroCream, MetroGel, MetroLotion, ✿Rosasol*): Rosacea: apply daily (1%) or bid (0.75%). [Trade only: Gel (MetroGel) 1% 45 g, Cream (Noritate) 1% 30, 60 g. Generic/Trade: Gel 0.75% 29g. Cream 0.75% 45 g. Lotion (MetroLotion) 0.75% 59 mL.] ▸KL ♀B(- in 1st trimester) ▸- $$$

mupirocin (*Bactroban, Centany*): Impetigo/infected wounds: apply tid. Nasal methicillin-resistant Staph aureus eradication: 0.5 g in each nostril bid x 5 days. [Generic/Trade: cream/ointment 2% 15, 22, 30 g, 2% nasal ointment 1 g single-use tubes (for MRSA eradication).] ▸Not absorbed ♀B ▸? $$$

Neosporin cream (neomycin + polymyxin): apply daily-tid. [OTC Trade only: neomycin 3.5 mg/g + polymyxin 10,000 units/g 15 g & unit dose 0.94 g.] ▸K ♀C ▸? $

Neosporin ointment (bacitracin + neomycin + polymyxin): apply daily-tid. [OTC Generic/Trade: bacitracin 400 units/g + neomycin 3.5 mg/g + polymyxin 5,000 units/g 2.4,9.6,14.2,15, 30 g and unit dose 0.94 g.] ▸K ♀C ▸? $

Polysporin (bacitracin + polymyxin, *✿Polytopic*): apply ointment/aerosol/powder daily-tid. [OTC Trade only: ointment 15,30 g and unit dose 0.9 g, powder 10 g, aerosol 90 g.] ▸K ♀C ▸? $

silver sulfadiazine (*Silvadene, ✿Dermazin, Flamazine, SSD*): apply daily-bid. [Generic/Trade: cream 1% 20,50,85,400,1000g.] ▸LK ♀B ▸- $

Antifungals

butenafine (*Lotrimin Ultra, Mentax*): apply daily-bid. [Trade only. Rx: cream 1% 15,30 g (Mentax). OTC: cream 1% (Lotrimin Ultra).] ▸L ♀B ▸? $$$

ciclopirox (**Loprox, Penlac, ♣Stieprox shampoo**): Cream, lotion: apply bid. Nail solution: apply daily to affected nails; apply over previous coat; remove with alcohol every 7 days. Seborrheic dermatitis (Loprox shampoo): shampoo twice/week x 4 weeks. [Trade only: gel (Loprox) 0.77% 30,45 g, shampoo (Loprox) 1% 120 mL, nail solution (Penlac) 8% 6.6 mL. Generic/Trade: cream (Loprox) 0.77% 15,30,90 g, lotion (Loprox) 0.77% 30,60 mL.] ▶LK ♀B ▶? $$$$

clotrimazole - topical (**Lotrimin, Mycelex, ♣Canesten, Clotrimaderm**): apply bid. [Note that Lotrimin brand cream, lotion, solution are clotrimazole, while Lotrimin powders and liquid spray are miconazole. OTC & Rx generic/Trade: cream 1% 15, 30, 45, 90 g, solution 1% 10,30 mL. Trade only: lotion 1% 30 mL.] ▶L ♀B ▶? $

econazole (**Spectazole, ♣Ecostatin**): Tinea pedis, cruris, corporis, tinea versicolor: apply daily. Cutaneous candidiasis: apply bid. [Generic/Trade: cream 1% 15, 30, 85g.] ▶Not absorbed ♀C ▶? $$

ketoconazole - topical (**Nizoral, ♣Ketoderm**): Tinea/candidal infections: apply daily. Seborrheic dermatitis: apply cream daily-bid. Dandruff: apply 1% shampoo twice a week. Tinea versicolor: apply shampoo to affected area, leave on for 5 min, rinse. [Trade only: shampoo 1% (OTC), 2%, 120 mL. Generic/Trade: cream 2% 15,30,60 g.] ▶L ♀C ▶? $$

miconazole -topical (**Monistat-Derm, Micatin, Lotrimin**): Tinea, candida: apply bid. [Note that Lotrimin brand cream, lotion, solution are clotrimazole, while Lotrimin powders & liquid spray are miconazole. OTC generic: ointment 2% 29 g, spray 2% 105 mL, soln 2% 7.39, 30 mL. Generic/Trade: cream 2% 15,30,90 g, powder 2% 90 g, spray powder 2% 90,100 g, spray liquid 2% 105,113 mL.] ▶L ♀+ ▶? $

naftifine (**Naftin**): Tinea: apply daily (cream) or bid (gel). [Trade only: cream 1% 15,30,60 g, gel 1% 20,40, 60 g.] ▶LK ♀B ▶? $$$

nystatin - topical (**Mycostatin, ♣Nilstat, Nyaderm, Candistatin**): Candidiasis: apply bid-tid. [Generic/Trade: cream 100,000 units/g 15,30,240 g, ointment 100,000 units/g 15,30 g, powder 100,000 units/g 15 g.] ▶Not absorbed ♀C ▶? $

oxiconazole (**Oxistat, Oxizole**): Tinea pedis, cruris, and corporis: apply daily-bid. Tinea versicolor (cream only): apply daily. [Trade only: cream 1% 15, 30, 60 g, lotion 1% 30 mL.] ▶? ♀B ▶? $$$

sertaconazole (**Ertaczo**): Tinea pedis: apply bid. [Trade only: cream 2% 15, 30, 60 g.] ▶Not absorbed ♀C ▶? $$

terbinafine (**Lamisil, Lamisil AT**): Tinea: apply daily-bid. [Trade only: cream 1% 15,30 g, gel 1% 5,15,30 g OTC: Trade only (Lamisil AT): cream 1% 12,24 g, spray pump solution 1% 30 mL.] ▶L ♀B ▶? $$$

tolnaftate (**Tinactin, ♣ZeaSorb AF**): apply bid. [OTC Generic/Trade: cream 1% 15,30 g, solution 1% 10,15 mL, powder 1% 45,90 g, spray powder 1% 100,105,150 g, spray liquid 1% 60,120 mL. Trade only: gel 1% 15 g.] ▶? ♀? ▶? $

Antiparasitics (topical)

A-200 (pyrethrins + piperonyl butoxide, **♣R&C**): Lice: Apply shampoo, wash after 10 min. Reapply in 5-7 days. [OTC Generic/Trade: shampoo (0.33% pyrethrins, 4% piperonyl butoxide) 60,120,240 mL.] ▶L ♀C ▶? $

crotamiton (**Eurax**): Scabies: apply cream/lotion topically from chin to feet, repeat in 24 hours, bathe 48 h later. Pruritus: massage prn. [Trade only: cream 10% 60 g, lotion 10% 60,480 mL.] ▶? ♀C ▶? $$

lindane (**♣Hexit**): Other drugs preferred. Scabies: apply 30-60 mL of lotion, wash after 8-12h. Lice: 30-60 mL of shampoo, wash off after 4 min. Can cause seizures in epileptics or if overused/misused in children. Not for infants. [Generic/Trade: lotion 1% 30,60,480 mL, shampoo 1% 30,60,480 mL.] ▶L ♀B ▶? $

malathion (*Ovide*): apply to dry hair, let dry naturally, wash off in 8-12 hrs. [Trade only: lotion 0.5% 59 mL.] ▶? ♀B ▶? $$

permethrin (*Elimite, Acticin, Nix, ✿Kwellada-P*): Scabies: apply cream from head (avoid mouth/ nose/eyes) to soles of feet & wash after 8-14h. 30 g is typical adult dose. Lice: Saturate hair and scalp with 1% rinse, wash after 10 min. Do not use in children <2 months old. May repeat therapy in 7 days, as necessary. [Trade only: cream (Elimite, Acticin) 5% 60 g. OTC Generic/Trade: liquid creme rinse (Nix) 1% 60 mL.] ▶L ♀B ▶? $$

RID (pyrethrins + piperonyl butoxide): Lice: Apply shampoo/mousse, wash after 10 min. Reapply in 5-10 days. [OTC Generic/Trade: shampoo 60,120,240 mL. Trade only: mousse 5.5 oz.] ▶L ♀C ▶? $

Antipsoriatics

acitretin (*Soriatane*): 25-50 mg PO daily. Avoid pregnancy during therapy and for 3 years after discontinuation. [Trade only: cap 10,25 mg.] ▶L ♀X ▶- $$$$$

alefacept (*Amevive*): 7.5 mg IV or 15 mg IM once weekly x 12 doses. May repeat with 1 additional 12-week course after 12 weeks have elapsed since last dose. ▶? ♀B ▶? $$$$$

anthralin (*Anthra-Derm, Drithocreme, ✿Anthrascalp, Anthranol, Anthraforte, Dithranol*): Apply daily. Short contact periods (i.e. 15-20 minutes) followed by removal may be preferred. [Trade only: ointment 0.1% 42.5 g, 0.25% 42.5 g, 0.4% 60 g, 0.5% 42.5 g, 1% 42.5 g, cream 0.1% 50 g, 0.2% 50 g, 0.25% 50 g, 0.5% 50 g. Generic/Trade: cream 1% 50 g.] ▶? ♀C ▶- $$

calcipotriene (*Dovonex*): apply bid. [Trade only: ointment 0.005% 30,60,100 g, cream 0.005% 30,60,100 g, scalp solution 0.005% 60 mL.] ▶L ♀C ▶? $$$

efalizumab (*Raptiva*): 0.7 mg/kg SC x 1 then 1 mg/kg SC q week. [Trade only: single use vials, 125 mg.] ▶L ♀C ▶? $$$$$

Taclonex (calcipotriene + betamethasone): Apply daily for up to 4 weeks. [Trade only: ointment calcipotriene 0.005% + betamethasone dipropionate 0.064% ointment 15, 30, 60g.] ▶L ♀C ▶? $$$$$

Antivirals (topical)

acyclovir - topical (*Zovirax*): Herpes genitalis: apply oint q3h (6 times/d) x 7 days. Recurrent herpes labialis: apply cream 5 times/day for 4 days. [Trade only: ointment 5% 15 g, cream 5% (2 & 5 g).] ▶K ♀C ▶? $$$$$

docosanol (*Abreva*): oral-facial herpes simplex: apply 5x/day until healed. [OTC Trade only: cream 10% 2 g.] ▶Not absorbed ♀B ▶? $

imiquimod (*Aldara*): Genital/perianal warts: apply 3 times weekly during sleeping hours for up to 16 weeks. Wash off after 8 hours. Nonhyperkeratotic, nonhypertrophic actinic keratoses on face/scalp in immunocompetent adults: apply 2 times weekly during sleeping hours for 16 weeks. Wash off after 8 hours. Primary superficial basal cell carcinoma: apply 5 times weekly x 6 weeks. Wash off after 8h. [Trade only: cream 5% 250 mg single use packets.] ▶Not absorbed ♀C ▶? $$$$$

penciclovir (*Denavir*): Herpes labialis: apply cream q2h while awake x 4 days. [Trade only: cream 1% 2 g tubes.] ▶Not absorbed ♀B ▶? $$$

podofilox (*Condylox, ✿Condyline, Wartec*): External genital warts (gel and solution) and perianal warts (gel only): apply bid for 3 consecutive days of a week and repeat for up to 4 wks. [Trade only: gel 0.5% 3.5 g, soln 0.5% 3.5 mL.] ▶? ♀C ▶? $$$

podophyllin (*Podocon-25, Podofin, Podofilm*): Warts: apply by physician. Not to be dispensed to patients. [Trade only: liquid 25% 15 mL.] ▶? ♀- ▶- $$

CORTICOSTEROIDS – TOPICAL*

	Agent	Strength/Formulation*	Freq
Low Potency	alclometasone dipropionate *(Aclovate)*	0.05% **C** O	bid-tid
	clocortolone pivalate *(Cloderm)*	0.1% **C**	tid
	desonide *(DesOwen, Tridesilon)*	0.05% **C L O**	bid-tid
	hydrocortisone *(Hytone)*, others	0.5% **C L O**; 1% **C L O**; 2.5% **C L O**	bid-qid
	hydrocortisone acetate *(Cortaid, Corticaine)*	0.5% **C** O, 1% **C** O SP	bid-qid
Medium Potency	betamethasone valerate *(Luxiq)*	0.1% **C L O**; 0.12% F*(Luxiq)*	qd-bid
	desoximetasone‡ *(Topicort)*	0.05% **C**	bid
	fluocinolone *(Synalar)*	0.01% **C** S; 0.025% **C** O	bid-qid
	flurandrenolide *(Cordran)*	0.025% CO; 0.05% CLO; T	bid-qid
	fluticasone propionate *(Cutivate)*	0.005% **O**; 0.05% **C** L	qd-bid
	hydrocortisone butyrate *(Locoid)*	0.1% **C** O S	bid-tid
	hydrocortisone valerate *(Westcort)*	0.2% **C** O	bid-tid
	mometasone furoate *(Elocon)*	0.1% **C L O**	qd
	triamcinolone‡ *(Aristocort, Kenalog)*	0.025% **CLO**; 0.1% **CLO**; SP	bid-tid
High Potency	amcinonide *(Cyclocort)*	0.1% **C L O**	bid-tid
	betamethasone dipropionate‡ *(Maxivate, others)*	0.05% **C L O** (non-Diprolene)	qd-bid
	desoximetasone‡ *(Topicort)*	0.05% **G**; 0.25% **C** O	bid
	diflorasone diacetate‡ *(Maxiflor)*	0.05% **C** O	bid
	fluocinonide *(Lidex)*	0.05% **C G O S**	bid-qid
	halcinonide *(Halog)*	0.1% **C** O S	bid-tid
	triamcinolone‡ *(Aristocort, Kenalog)*	0.5% **C** O	bid-tid
Very high	betamethasone dipropionate‡ *(Diprolene, Diprolene AF)*	0.05% **C G L O**	qd-bid
	clobetasol *(Temovate, Cormax, Olux)*	0.05% **C G O L S** SP F *(Olux)*	bid
	diflorasone diacetate‡ *(Psorcon)*	0.05% **C** O	qd-tid
	halobetasol propionate *(Ultravate)*	0.05% **C** O	qd-bid

*C-cream, G-gel, L-lotion, O-ointment, S-solution, T-tape, F-foam, SP-spray; **bolded** items have available generics. Potency based on vasoconstrictive assays, which may not correlate with efficacy. Not all available products are listed, including those lacking potency ratings. ‡These drugs have formulations in more than once potency category.

Atopic Dermatitis Preparations

pimecrolimus (***Elidel***): Atopic dermatitis: apply bid. [Trade only: cream 1% 30, 60, 100 g.] ▶L ♀C ▶? $$$$

tacrolimus - topical (***Protopic***): Atopic dermatitis: apply bid. [Trade only: ointment 0.03% 30, 60, 100 g, 0.1% 30, 60, 100 g.] ▶Minimal absorption ♀C ▶? $$$$

Corticosteroid / Antimicrobial Combinations

Cortisporin (neomycin + polymyxin + hydrocortisone): apply bid-qid. [Generic/Trade: cream 7.5 g, ointment 15 g.] ▶LK ♀C ▶? $

Fucidin H (fusidic acid + hydrocortisone): Canada only. apply tid. [Trade only: cream 2% fusidic acid, 1% hydrocortisone acetate 30 g.] ▶L ♀? ▶? $

Lotrisone (clotrimazole + betamethasone, ✦*Lotriderm*): Apply bid. Do not use for diaper rash. [Generic/Trade: cream (clotrimazole 1% + betamethasone 0.05%) 15,45 g, lotion (clotrimazole 1% + betamethasone 0.05%) 30 mL.] ▶L ♀C ▶? $$$

Mycolog II (nystatin + triamcinolone): apply bid. [Generic/Trade: cream 15,30,60, 120 g, ointment 15,30,60,120 g.] ▶L ♀C ▶? $

Hemorrhoid Care

dibucaine (**Nupercainal**): apply cream/ointment tid-qid prn. [OTC Generic/Trade: ointment 1% 30 g.] ▶L ♀? ▶? $

hydrocortisone - topical (**Anusol-HC, Proctocream HC, Cortifoam, ✦Hemcort HC, Egozinc-HC**): apply cream tid-qid prn or supp bid or foam daily-bid. [Generic/Trade: cream 1% (Anusol HC-1), 2.5% 30 g (Anusol HC, Proctocream HC), suppositories 25 mg (Anusol HC), 10% rectal foam (Cortifoam) 15g.] ▶L ♀? ▶? $

pramoxine (**Anusol Hemorrhoidal Ointment, Fleet Pain Relief, Proctofoam NS**): ointment/pads/foam up to 5 times/day prn. [OTC Trade only: ointment (Anusol Hemorrhoidal Ointment), pads (Fleet Pain Relief), aerosol foam (ProctoFoam NS).] ▶Not absorbed ♀+ ▶+ $

starch (**Anusol Suppositories**): 1 suppository up to 6 times/day prn. [OTC Trade only: suppositories (51% topical starch; soy bean oil, tocopheryl acetate).] ▶Not absorbed ♀+ ▶+ $

witch hazel (**Tucks**): Apply to anus/perineum up to 6 times/day prn. [OTC Generic/Trade: pads, gel.] ▶? ♀+ ▶+ $

Other Dermatologic Agents

alitretinoin (**Panretin**): Apply bid-qid to cutaneous Kaposi's lesions [Trade only: gel 60 g.] ▶Not absorbed ♀D ▶- $$$$$

aluminum chloride (**Drysol, Certain Dri**): Apply qhs. [Rx: Generic/Trade: solution 20%: 37.5 mL bottle, 35, 60 mL bottle with applicator. OTC: Trade only (Certain Dri): solution 12.5%: 36 mL bottle.] ▶K ♀? ▶? $

becaplermin (**Regranex**): Diabetic ulcers: apply gel daily. [Trade only: gel 0.01% 2, 15 g.] ▶Minimal absorption ♀C ▶? $$$$$

calamine: apply lotion tid-qid prn for poison ivy/oak or insect bite itching. [OTC Generic: lotion 120, 240, 480 mL.] ▶? ♀? ▶? $

capsaicin (**Zostrix, Zostrix-HP**): Arthritis, post-herpetic or diabetic neuralgia: apply cream up to tid-qid. [OTC Generic/Trade: cream 0.025% 45,60 g, 0.075% 30,60 g, lotion 0.025% 59 mL, 0.075% 59 mL, gel 0.025% 15,30, 0.05% 43 g, roll-on 0.075% 60 mL.] ▶? ♀? ▶? $

coal tar (**Polytar, Tegrin, Cutar, Tarsum**): shampoo at least twice a week, or for psoriasis apply daily-qid. [OTC Generic/Trade: shampoo, conditioner, cream, ointment, gel, lotion, soap, oil.] ▶? ♀? ▶? $

doxepin - topical (**Zonalon**): Pruritus: apply qid for up to 8 days. [Trade only: cream 5% 30,45 g.] ▶L ♀B ▶- $

eflornithine (**Vaniqa**): Reduction of facial hair: apply to face bid. [Trade only: cream 13.9% 30 g.] ▶K ♀C ▶? $$

EMLA (lidocaine + prilocaine): Topical anesthesia: apply 2.5g cream or 1 disc to region at least 1 hour before procedure. Cover cream with an occlusive dressing. [Trade only: cream (2.5% lidocaine + 2.5% prilocaine) 5 g, disc 1 g. Generic/Trade: cream (2.5% lidocaine + 2.5% prilocaine) 30 g.] ▶LK ♀B ▶? $$

finasteride - topical (**Propecia**): Androgenetic alopecia in men: 1 mg PO daily. [Generic/Trade: tab 1 mg.] ▶L ♀X ▶- $$$

hyaluronic acid (**Restylane**): Moderate to severe facial wrinkles: inject into wrinkle/fold. [Rx: 0.4 mL and 0.7 mL syringe.] ▶? ♀? ▶? $$$$$

hydroquinone (**Eldopaque, Eldoquin, Eldoquin Forte, EpiQuin Micro, Esoterica, Glyquin, Lustra, Melanex, Solaquin, Claripel, ✦Ultraquin, Neostrata HQ**): Hyperpigmentation: apply to area bid. [OTC Generic/Trade: cream 1.5%, lotion 2%. Rx Generic/Trade: solution 3%, gel 4%, cream 4%.] ▶? ♀C ▶? $$$

lactic acid (**Lac-Hydrin, Amlactin**): apply lotion/cream bid. [Trade: lotion 12% 150,360 mL. Generic/OTC: cream 12% 140,385 g. AmLactin AP is lactic acid (12%) with pramoxine (1%).] ▶? ♀? ▶? $$

lidocaine - topical (**Xylocaine, Lidoderm, Numby Stuff, LMX, ✦Maxilene**): Apply prn. Dose varies with anesthetic procedure, degree of anesthesia required and individual response. Postherpetic neuralgia: Postherpetic neuralgia: apply up to 3 patches to affected area at once for up to 12h within a 24h period. Apply 30 min prior to painful procedure (ELA-Max 4%). Discomfort with anorectal disorders: apply prn (ELA-Max 5%). [For mouth and pharynx: spray 10%, ointment 5%, liquid 5%, solution 2,4%, dental patch. For urethral use: jelly 2%. Patch (Lidoderm) 5%. OTC: Trade only: liposomal lidocaine 4% (ELA-Max).] ▶LK ♀B ▶+ $$

methylaminolevulinate (**Metvixia**): Apply cream to non-hyperkeratotic actinic keratoses lesion and surrounding area on face or scalp; cover with dressing for 3h; remove dressing and cream and perform illumination therapy. Repeat in 7 days. [Trade only: cream: 16.8%, 2 g.] ▶Not absorbed ♀C ▶? ?

minoxidil - topical (**Rogaine, Rogaine Forte, Rogaine Extra Strength, Minoxidil for Men, Theroxidil Extra Strength, ✦Minox, Apo-Gain**): Androgenetic alopecia in men or women: 1 mL to dry scalp bid. [OTC Trade only: solution 2%, 60 mL, 5%, 60 mL, 5% (Rogaine Extra Strength, Theroxidil Extra Strength - for men only) 60 mL, foam 5% 60g.] ▶K ♀C ▶- $$

monobenzone (**Benoquin**): Extensive vitiligo: apply bid-tid. [Trade only: cream 20% 35.4 g.] ▶Minimal absorption ♀C ▶? $$

oatmeal (**Aveeno**): Pruritus from poison ivy/oak, varicella: apply lotion qid prn. Also bath packets for tub. [OTC Generic/Trade: lotion, packets.] ▶Not absorbed?♀? ▶? $

Panafil (papain + urea + chlorophyllin copper complex): Debridement of acute or chronic lesions: apply to clean wound and cover daily-bid. [Trade only: Ointment: 6, 30 g, spray 33 mL.] ▶? ♀? ▶? $$$

Pramosone (pramoxine + hydrocortisone, ✦**Pramox HC**): Inflammatory and pruritic manifestations of corticosteroid-responsive dermatoses: Apply tid-qid. [Trade only: 1% pramoxine/1% hydrocortisone acetate: cream 30, 60 g, oint 30 g, lotion 60, 120, 240 mL. 1% pramoxine/2.5% hydrocortisone acetate: cream 30, 60 g, oint 30 g, lotion 60, 120 mL.] ▶Not absorbed ♀C ▶? $$$

selenium sulfide (**Selsun, Exsel, Versel**): Dandruff, seborrheic dermatitis: apply 5-10 mL lotion/shampoo twice weekly x 2 weeks then less frequently, thereafter. Tinea versicolor: Apply 2.5% lotion/shampoo to affected area daily x 7 days. [OTC Generic/Trade: lotion/shampoo 1% 120,210,240, 330 mL, 2.5% 120 mL. Rx Generic/Trade: lotion/shampoo 2.5% 120 mL.] ▶? ♀C ▶? $

Solag (mequinol + tretinoin, ✦**Solagé**): Apply to solar lentigines bid. [Trade only: soln 30 mL (mequinol 2% + tretinoin 0.01%).] ▶Not absorbed ♀X ▶? $$$$$

Synera (lidocaine + tetracaine): Apply 20-30 min prior to superficial derm procedure. [Trade only: patch lido 70 mg + tetra 70 mg.] ▶Minimal absorption ♀B ▶?

Tri-Luma (fluocinolone + hydroquinone + tretinoin): Melasma of the face: apply qhs x 4-8 weeks. [Trade only: soln 30 g (fluocinolone 0.01% + hydroquinone 4% + tretinoin 0.05%).] ▶Minimal absorption ♀C ▶? $$$

Vusion (miconazole + zinc oxide + white petrolatum): Apply to affected diaper area with each change for 7 days. [Trade only: 30 g.] ▶Minimal absorption ♀C ▶? $$$

ENDOCRINE & METABOLIC

Androgens / Anabolic Steroids (See OB/GYN section for other hormones.)

methyltestosterone (**Android, Methitest, Testred, Virilon**): Advancing inoperable breast cancer in women who are 1-5 years postmenopausal: 50-200 mg/day PO in divided doses. Hypogonadism in men: 10-50 mg PO daily. [Trade only: Caps 10 mg. Generic/Trade: Tabs 10, 25 mg.] ▶L ♀X ▶? ©III $$$

nandrolone (**Deca-Durabolin**): Anemia of renal disease: women 50-100 mg IM q week, men 100-200 mg IM q week. [Generic/Trade: injection 100 mg/mL, 200 mg/mL. Trade only: injection 50 mg/mL.] ▶L ♀X ▶- ©III $$$

oxandrolone (**Oxandrin**): Weight gain: 2.5 mg PO bid-qid for 2-4 wks. [Trade only: Tabs 2.5, 10 mg.] ▶L ♀X ▶? ©III $$$$$

testosterone (**Androderm, AndroGel, Delatestryl, Depo-Testosterone, Depotest, Everone 200, Striant, Testim, Testopel, Testro AQ, Testro-L.A., Virilon IM, ♣Andriol**): Injectable enanthate or cypionate: 50-400 mg IM q2-4 wks. Transdermal - Androderm: 5 mg patch to nonscrotal skin qhs. AndroGel 1%: Apply 5 g from gel pack or 4 pumps (5 g) from dispenser daily to shoulders/upper arms/abdomen. Testim: One tube (5 g) daily to shoulders/upper arms. Pellet - Testopel: 2-6 (150-450 mg testosterone) pellets SC q 3-6 months. Buccal- Striant: 30 mg q 12 hours on upper gum above the incisor tooth; alternate sides for each application. [Trade only: Patch 2.5, 5 mg (Androderm). 75 g multi-dose pump (AndroGel). Gel 1%, 5 g tube (Testim). Pellet 75mg (Testopel). Buccal: blister packs - 30 mg (Striant). Injection 200 mg/mL (enanthate). Generic/Trade: Gel 1% 2.5,5g foil packet. Injection 100, 200 mg/mL (cypionate).] ▶L ♀X ▶? ©III $$$$

Bisphosphonates

alendronate (**Fosamax, Fosamax Plus D**): Postmenopausal osteoporosis prevention (5 mg PO daily or 35 mg PO weekly) & treatment (10 mg daily, 70 mg PO weekly or 70 mg/vit D3 2800 IU PO weekly). Treatment of glucocorticoid-induced osteoporosis in men & women: 5 mg PO daily or 10 mg PO daily (postmenopausal women not taking estrogen). Treatment of osteoporosis in men: 10 mg PO daily, 70 mg PO weekly, or 70 mg/vit D3 2800 IU PO weekly. Paget's disease in men & women: 40 mg PO daily x 6 mon. May cause severe esophagitis. [Trade only: Tabs 5,10,35,40,70 mg, 70mg/Vit D3 2800 I.U. (Fosamax Plus D). Oral soln 70 mg/75 mL (single dose bottle).] ▶K ♀C ▶- $$$

clodronate (**♣Ostac, Bonefos**): Canada only. IV single dose - 1500 mg slow infusion over ≥4 hours. IV multiple dose - 300 mg slow infusion daily over 2-6 hours up to 10 days. Oral - following IV therapy, maintenance 1600-2400 mg/day in single or divided doses. Max PO dose 3200 mg/day; duration of therapy is usually 6 months. [Trade only: Capsules 400 mg.] ▶K ♀D ▶- $$$$

etidronate (**Didronel**): Hypercalcemia: 7.5 mg/kg in 250 mL NS IV over ≥2h daily x 3d. Paget's disease: 5-10 mg/kg PO daily x 6 months or 11-20 mg/kg daily x 3 months. [Generic/Trade: Tabs 200, 400 mg.] ▶K ♀C ▶? $$$$

ibandronate (**Boniva**): Treatment/Prevention of postmenopausal osteoporosis. Oral: 2.5 mg PO daily or 150 mg PO q month. IV: 3 mg IV every 3 months. [Trade only: 2.5, 150 mg tabs.] ▶K ♀C ▶? $$$

pamidronate (**Aredia**): Hypercalcemia of malignancy: 60-90 mg IV over 2-24 h. Wait ≥7 days before considering retreatment. ▶K ♀D ▶? $$$$$

risedronate (**Actonel, Actonel Plus Calcium**): Paget's disease: 30 mg PO daily x 2 months. Prevention & treatment of postmenopausal osteoporosis: 5 mg PO daily or 35 mg PO weekly. Prevention & treatment of glucocorticoid-induced osteoporosis: 5 mg PO daily. May cause esophagitis. [Trade only: Tabs 5, 30, 35 mg, 35/1250 mg (calcium).] ▶L ♀C ▶? $$$

zoledronic acid (**Zometa, ♣Aclasta**): Hypercalcemia: 4 mg IV infusion over ≥15 min. Wait ≥7 days before considering retreatment. Multiple myeloma and metastatic bone lesions from solid tumors: 4 mg IV infusion over ≥15 min q3-4 weeks. Osteoporosis: 4 mg IV dose once yearly. Paget's Disease: 5 mg IV single dose. ▶K ♀D ▶? $$$$$

Corticosteroids (See also dermatology, ophthalmology.)

betamethasone (**Celestone, Celestone Soluspan, ♣Betnesol, Betaject**): Anti-inflammatory/Immunosuppressive: 0.6-7.2 mg/day PO divided bid-qid; up to 9 mg/day IM. Fetal lung maturation, maternal antepartum: 12 mg IM q24h x 2 doses. [Trade only: Syrup 0.6 mg/5 mL.] ▶L ♀C ▶- $$$$$

cortisone (**Cortone**): 25-300 mg PO daily. [Generic/Trade: Tabs 5, 10, 25 mg.] ▶L ♀D ▶- $

dexamethasone (**Decadron, Dexone, Dexpak, ♣Dexasone, Maxidex**): Anti-inflammatory/Immunosuppressive: 0.5-9 mg/day PO/IV/IM, divided bid-qid. Fetal lung maturation, maternal antepartum: 6 mg IM q12h x 4 doses. [Generic/Trade: Tabs 0.25, 0.5, 0.75, 1.0, 1.5, 2, 4, 6 mg. elixir/ solution 0.5 mg/5 mL. Trade only: oral solution 0.5 mg/ 0.5 mL. Decadron Unipak (0.75 mg-12 tabs). Dexpak (51 total 1.5 mg tabs for a 13 day taper).] ▶L ♀C ▶- $

fludrocortisone (**Florinef**): Mineralocorticoid activity: 0.1 mg PO 3 times weekly to 0.2 mg PO daily. [Generic/Trade: Tabs 0.1 mg.] ▶L ♀C ▶? $

hydrocortisone (**Cortef, Cortenema, Hydrocortone, Solu-Cortef**): 100-500 mg IV/IM q2-6h prn (sodium succinate). 20-240 mg/day PO divided tid- qid. Ulcerative colitis: 100 mg retention enema qhs (laying on side for ≥1 hour) for 21 days. [Trade only: Tabs 5 mg. Generic/Trade: Tabs 10, 20 mg. Enema 100 mg/60 mL.] ▶L ♀C ▶- $

methylprednisolone (**Solu-Medrol, Medrol, Depo-Medrol**): Oral (Medrol): dose varies, 4-48 mg PO daily. Medrol Dosepak tapers 24 to 0 mg PO over 7 days. IM/Joints (Depo-Medrol): dose varies, 4-120 mg IM q1-2 weeks. Parenteral (Solu-Medrol): dose varies, 10-250 mg IV/IM. Acute spinal cord injury: 30 mg/kg IV over 15 min, followed in 45 min by a 5.4 mg/kg/hr IV infusion x 23-47h. Peds: 0.5-1.7 mg/kg PO/IV/IM divided q6-12h. [Trade only: Tabs 2, 8, 16, 24, 32 mg. Generic/Trade: Tabs 4 mg. Medrol Dosepak (4 mg-21 tabs).] ▶L ♀C ▶- $$$

prednisolone (**Delta-Cortef, Prelone, Pediapred, Orapred, Orapred ODT**): 5-60 mg PO/IV/IM daily. [Generic/Trade: Tabs 5 mg; orally disintegrating tabs: 10, 15, 30 mg (Orapred ODT); syrup 5 mg/5 mL, 15 mg/5 mL (Prelone; wild cherry flavor). Generic/Trade: solution 5 mg/5 mL (Pediapred; raspberry flavor); solution 15 mg/5 mL (Orapred; grape flavor).] ▶L ♀C ▶+ $$

prednisone (**Deltasone, Meticorten, Pred-Pak, Sterapred, ♣Winpred**): 1-2 mg/kg or 5-60 mg PO daily. [Generic/Trade: Tabs 1, 5, 10, 20, 50 mg. Trade only: Tabs 2.5 mg. Sterapred (5 mg tabs: tapers 30 to 5 mg PO over 6d or 30 to 10 mg over 12d), Sterapred DS (10 mg tabs: tapers 60 to 10 mg over 6d, or 60 to 20 mg PO over 12d) & Pred-Pak (5 mg 45 & 79 tabs) taper packs. Solution 5 mg/5 mL & 5 mg/mL (Prednisone Intensol).] ▶L ♀C ▶+ $

triamcinolone (**Aristocort, Kenalog, ♣Aristospan**): 4-48 mg PO/IM daily. [Generic/Trade: Tabs 4 mg. Trade only: injection 10 mg/mL, 25 mg/mL, 40 mg/mL.] ▶L ♀C ▶- $$$$

CORTICOSTEROIDS	Approximate equivalent dose (mg)	Relative anti-inflammatory potency	Relative mineralocorticoid potency	Biologic Half-life (hours)
betamethasone	0.6-0.75	20-30	0	36-54
cortisone	25	0.8	2	8-12
dexamethasone	0.75	20-30	0	36-54
fludrocortisone	--	10	125	18-36
hydrocortisone	20	1	2	8-12
methylprednisolone	4	5	0	18-36
prednisolone	5	4	1	18-36
prednisone	5	4	1	18-36
triamcinolone	4	5	0	12-36

Diabetes-Related - Alphaglucosidase Inhibitors

acarbose (Precose, ♥Prandase): Start 25 mg PO tid with meals, and gradually increase as tolerated to maintenance 50-100 mg tid. [Trade only: Tabs 25, 50, 100 mg.] ▶Gut/K ♀B ▶- $$

miglitol (Glyset): Start 25 mg PO tid with meals, maintenance 50-100 tid. [Trade only: Tabs 25, 50, 100 mg.] ▶K ♀B ▶- $$$

Diabetes-Related - Combinations

ACTOPLUS Met (pioglitazone + metformin): 1 tablet PO daily-bid. If inadequate control with metformin monotherapy, select tab strength based on adding 15 mg pioglitazone to existing metformin dose. If inadequate control with pioglitazone monotherapy, select tab strength based on initiating 500 mg bid or 850 mg daily metformin to existing pioglitazone dose. Max 45/2550 mg/day. [Trade only: Tabs 15/500, 15/850 mg.] ▶KL ♀C ▶? $$$$

Avandamet (rosiglitazone + metformin): As initial therapy: Start 2/500 mg PO daily or bid. As second-line therapy: 1 tab PO bid. If inadequate control with metformin monotherapy, select tab strength based on adding 4 mg/day rosiglitazone to existing metformin dose. If inadequate control with rosiglitazone monotherapy, select tab strength based on adding 1000 mg/day metformin to existing rosiglitazone dose. Max 8/2000 mg/day. Obtain LFTs before therapy & periodically thereafter. [Trade only: Tabs 1/500, 2/500, 4/500, 2/1000, 4/1000 mg.] ▶KL ♀C ▶? $$$$

Avandaryl (rosiglitazone + glimepiride): Type 2 DM inadequately controlled on rosiglitazone or sulfonylurea alone: Start 4/1 or 4/2 mg PO daily or divided bid, max 8/4 mg per day. Obtain LFTs before therapy & periodically thereafter. [Trade only: Tabs 4/1, 4/2, 4/4 mg rosiglitazone/glimepiride.] ▶LK ♀C ▶? $$$$

Duetact (pioglitazone + glimepiride): Type 2 DM inadequately controlled on pioglitazone or sulfonylurea alone: Start 30/2 mg PO daily. May start 30/4 mg PO daily if prior glimepiride therapy, max 30/4 mg per day. Obtain LFTs before therapy & periodically thereafter. [Trade only: Tabs 30/2, 30/4 mg pioglitazone/glimepiride.] ▶LK ♀C ▶- $$$$

Glucovance (glyburide + metformin): As initial therapy: Start 1.25/250 mg PO daily or bid with meals; max 10/2000 mg daily. As second-line therapy: Start 2.5/500 or 5/500 mg PO bid with meals; max 20/2000 mg daily. [Generic/Trade: Tabs 1.25/250, 2.5/500, 5/500 mg] ▶KL ♀B ▶? $$$

Metaglip (glipizide + metformin): As initial therapy: Start 2.5/250 mg PO daily to 2.5/500 mg PO bid with meals; max 10/2000 mg daily. As second-line therapy: Start 2.5/500 or 5/500 mg PO bid with meals; max 20/2000 mg daily. [Generic/Trade: Tabs 2.5/250, 2.5/500, 5/500 mg.] ▶KL ♀C ▶? $$$

Diabetes-Related - "Glitazones" (Thiazolidinediones)

pioglitazone (**Actos**): Start 15-30 mg PO daily, max 45 mg/day. Monitor LFTs. [Trade only: Tabs 15, 30, 45 mg.] ▶L ♀C ▶- $$$$

rosiglitazone (**Avandia**): Diabetes monotherapy or in combination with metformin, sulfonylurea or insulin: Start 4 mg PO daily or divided bid, max 8 mg/day monotherapy or in combination with metformin and/or sulfonylurea. Max 4 mg/day when used with insulin. Obtain LFTs before therapy & periodically thereafter. [Trade only: Tabs 2, 4, 8 mg.] ▶L ♀C ▶- $$$$

Diabetes-Related - Insulins

insulin, inhaled (**Exubera**): Initial per meal dose based on patient body weight. 1 mg (30-39.9 kg); 2 mg (40-59.9 kg); 3 mg (60-79.9 kg); 4 mg (80-99.9 kg); 5 mg (100-119.9 kg); 6 mg (120-139.9 kg). Rapid-acting; administer immediately before a meal (within 10 min). [Trade only: Combination patient pack 90 x 1 mg insulin blisters and 90 x 3 mg insulin blisters; 1 mg patient pack (90 x 1 mg insulin blisters); 3 mg patient pack (90 x 3 mg insulin blisters).] ▶LK ♀C ▶+ $$$$

insulin, injectable combinations (**Humalog Mix 75/25, Humalog Mix 50/50, Humulin 70/30, Humulin 50/50, Novolin 70/30, Novolog Mix 70/30**): Diabetes: Doses vary, but typically total insulin 0.3-0.5 unit/kg/day SC in divided doses (Type 1), and 1-1.5 unit/kg/day SC in divided doses (Type 2). Administer rapid-acting insulin mixtures (Humalog, NovoLog) within 15 min before. Administer regular insulin mixtures 30 minutes before meals. [Trade only: injection insulin lispro protamine suspension / insulin lispro (Humalog Mix 75/25, Humalog Mix 50/50), insulin aspart protamine/insulin aspart (Novolog Mix 70/30) NPH and regular mixtures (Humulin 70/30, Novolin 70/30 or Humulin 50/50). Insulin available in pen form: Novolin 70/30 Innolet, Novolog Mix 70/30 FlexPen, Humulin 70/30, Humalog Mix 75/25, Humalog Mix 50/50.] ▶LK ♀B/C ▶+ $$$

DIABETES NUMBERS*

Self-monitoring glucose goals	
Preprandial	90-130 mg/dL
Postprandial	< 180 mg/dL
A1C Normal/individualized goal <6%; general goal <7%	

Criteria for diagnosis (confirm on different day)
Pre-diabetes: Fasting glucose 100-125 mg/dL
Diabetes: Fasting glucose ≥126 mg/dL, or random glucose with symptoms: ≥200 mg/dL

Complications prevention & management: Aspirin** (75–162 mg/day) in Type 1 & 2 adults for primary prevention (those with an increased cardiovascular risk, including >40 yo or with additional risk factors) and secondary prevention (those with any vascular disease) unless contraindicated; statin therapy to achieve 30% LDL reduction for >40 yo & total cholesterol ≥135 mg/dL;† ACE inhibitor or ARB if hypertensive or micro-/macro-albuminuria; pneumococcal vaccine (revaccinate one time if age >64 and previously received vaccine at age <65 and > 5 years ago).

At every visit: Measure weight & BP (goal <130/80 mmHg); visual foot exam; review self-monitoring glucose record; review/adjust meds; review self-management skills, dietary needs, and physical activity; smoking cessation counseling.

Twice a year: A1C in those meeting treatment goals with stable glycemia (quarterly if not); dental exam.

Annually: Fasting lipid profile (goal LDL <100 mg/dL, cardiovascular disease consider LDL <70mg/dL, HDL >40 mg/dL, TG <150 mg/dL) q2 years with low-risk lipid values; creatinine; albumin to creatinine ratio spot collection; dilated eye exam; flu vaccine; comprehensive foot exam.

*See recommendations at: http://care.diabetesjournals.org. Glucose values are plasma.
**Avoid aspirin if <21 yo due to Reye's Syndrome risk; use if <30 yo has not been studied.
†ADA. *Diabetes Care* 2006;29 (Suppl 1):S4-S42, *Circulation.* 2004; 110:235

insulin, injectable intermediate/long-acting (**Novolin N, Humulin N, Lantus, Levemir**): Diabetes: Doses vary, but typically total insulin 0.3-0.5 unit/kg/day SC in divided doses (Type 1), and 1-1.5 unit/kg/day SC in divided doses (Type 2). Generally, 50-70% of insulin requirements are provided by rapid or short-acting insulin and the remainder from intermediate- or long-acting insulin. Lantus: Start 10 units SC daily (same time everyday) in insulin naive patients. Levemir: Type 2 DM (inadequately controlled on oral meds): Start 0.1-0.2 units/kg once daily in evening or 10 units SC daily or BID. [Trade only: injection NPH (Novolin N, Humulin N), insulin glargine (Lantus), insulin detemir (Levemir). Insulin available in pen form: Humulin N Isophane Pen, Novolin N Innolet, Lantus OptiClick, Levemir Innolet, Levemir FlexPen. Premixed preparations of NPH and regular insulin also available.] ▶LK ♀B/C ▶+ $$$

insulin, injectable short/rapid-acting (**Apidra, Novolin R, NovoLog, Humulin R, Humalog, ♣NovoRapid**): Diabetes: Doses vary, but typically total insulin 0.3-0.5 unit/kg/day SC in divided doses (Type 1), and 1-1.5 unit/kg/day SC in divided doses (Type 2). Generally, 50-70% of insulin requirements are provided by rapid or short-acting insulin and the remainder from intermediate- or long-acting insulin. Administer rapid-acting insulin (Humalog, NovoLog, Apidra) within 15 min before or immediately after a meal. Administer regular insulin 30 minutes before meals. Severe hyperkalemia: 5-10 units regular insulin plus concurrent dextrose IV. Profound hyperglycemia (eg, DKA): 0.1 unit regular/kg IV bolus, then initial infusion 100 units regular in 100 mL NS (1 unit/mL), at 0.1 units/kg/hr. 70 kg: 7 units/h (7 mL/h). [Trade only: injection regular, insulin glulisine (Apidra), insulin lispro (Humalog), insulin aspart (NovoLog). Insulin available in pen form: Novolog FlexPen, Humalog, Novolin R, Apidra OptiClik.] ▶LK ♀B/C ▶ $$$

INSULINS, INJECTABLE*		Onset (h)	Peak (h)	Duration (h)
Rapid/short-acting:	Insulin aspart (*NovoLog*)	<0.2	1-3	3-5
	Insulin glulisine (*Apidra*)	0.30-0.4	1	4-5
	Insulin lispro (*Humalog*)	0.25-0.5	0.5-2.5	≤ 5
	Regular (*Novolin R, Humulin R*)	0.5-1	2-3	3-6
Intermediate /long-acting:	NPH (*Novolin N, Humulin N*)	2-4	4-10	10-16
	Insulin detemir (*Levemir*)	not available	flat action profile	up to 23†
	Insulin glargine (*Lantus*)	2-4	peakless	24
Mixtures:	Insulin aspart protamine suspension/aspart (*NovoLog Mix 70/30*)	0.25	1-4 (biphasic)	up to 24
	Insulin lispro protamine suspension/insulin lispro (*HumaLog Mix 75/25, HumaLog Mix 50/50*)	<0.25	1-3 (biphasic)	10-20
	NPH/Reg (*Humulin 70/30, Humulin 50/50, Novolin 70/30*)	0.5-1	2-10 (biphasic)	10-20

*These are general guidelines, as onset, peak, and duration of activity are affected by the site of injection, physical activity, body temperature, and blood supply.
† Dose dependent duration of action, range from 6-23 hours.

Diabetes-Related - Meglitinides

nateglinide (**Starlix**): 120 mg PO tid ≤30 min before meals; use 60 mg PO tid in patients who are near goal A1C. [Trade only: Tabs 60, 120 mg.] ▶L ♀C ▶? $$$$

repaglinide (**Prandin, ♣Gluconorm**): Start 0.5- 2 mg PO tid before meals, maintenance 0.5-4 mg tid-qid, max 16 mg/day. [Trade: Tabs 0.5, 1, 2 mg] ▶L ♀C ▶? $$$

Diabetes-Related - Sulfonylureas

gliclazide (*♥Diamicron, Diamicron MR*): Canada only. Immediate release: Start 80-160 mg PO daily, max 320 mg PO daily (≥160 mg in divided doses). Modified release: Start 30 mg PO daily, max 120 mg PO daily. [Generic/Trade: Tab 80 mg (Diamicron). Trade only: Tab 30 mg (Diamicron MR).] ▶KL ♀C ▶? $

glimepiride (*Amaryl*): Start 1-2 mg PO daily, usual 1-4 mg/day, max 8 mg/day. [Generic/Trade: Tabs 1, 2, 4 mg. Generic only: Tabs 8 mg.] ▶LK ♀C ▶- $$

glipizide (*Glucotrol, Glucotrol XL*): Start 5 mg PO daily, usual 10-20 mg/day, max 40 mg/day (divide bid if >15 mg/day). Extended release: Start 5 mg PO daily, usual 5-10 mg/day, max 20 mg/day. [Generic/Trade: Tabs 5, 10 mg; Extended release tabs 2.5, 5, 10 mg.] ▶LK ♀C ▶? $

glyburide (*Micronase, DiaBeta, Glynase PresTab, ♥Euglucon*): Start 1.25-5 mg PO daily, usual 1.25-20 mg daily or divided bid, max 20 mg/day. Micronized tabs: Start 1.5-3 mg PO daily, usual 0.75-12 mg/day divided bid, max 12 mg/d. [Generic/Trade: Tabs (scored) 1.25, 2.5, 5 mg. micronized tabs (scored) 1.5, 3, 4.5, 6 mg.] ▶LK ♀B ▶? $

Diabetes-Related - Other

A1C home testing (*Metrika A1CNow*): For home A1C testing. ▶None ♀+ ▶+ $

dextrose (*Glutose, B-D Glucose, Insta-Glucose*): Hypoglycemia: 0.5-1 g/kg (1-2 mL/kg) up to 25 g (50 mL) of 50% soln IV. Dilute to 25% for pediatric administration. [OTC/Trade only: chew Tabs 5 g. gel 40%.] ▶L ♀C ▶? $

diazoxide (*Proglycem*): Hypoglycemia: 3-8 mg/kg/day divided q8-12h, max 10- 15 mg/kg/day. [Trade only: susp 50 mg/ mL.] ▶L ♀C ▶? $$$$$

exenatide (*Byetta*): Type 2 DM who are receiving metformin, sulfonylurea or both: 5 mcg SC bid (within 1 h before the morning and evening meals). May increase to 10 mcg SC bid after 1 month. [Trade only: prefilled pen (60 doses each) 5 mcg/dose, 1.2 mL; 10 mcg/dose, 2.4 mL.] ▶K ♀C ▶? $$$$

glucagon (*Glucagon, GlucaGen*): Hypoglycemia: 1 mg IV/IM/SC, onset 5-20 min. Diagnostic aid: 1 mg IV/IM/SC. [Trade only: injection 1 mg.] ▶LK ♀B ▶? $$$

glucose home testing (*Accu-Chek Active, Accu-Chek Advantage, Accu-Check Compact, Ascencia, FreeStyle, GlucoWatch, OneTouch InDuo, OneTouch Ultra, OneTouch UltraSmart, Precision Sof-Tact, Precision QID, Precision Xtra, True Track Smart System, Clinistix, Clinitest, Diastix, Tes-Tape, Gluco-Watch*): Use for home glucose monitoring. ▶None ♀+ ▶+ $$

metformin (*Glucophage, Glucophage XR, Glumetza, Fortamet, Riomet*): Diabetes: Start 500 mg PO daily-bid with meals, may gradually increase to max 2550 mg/day or 2000 mg/day (ext'd release). Polycystic ovary syndrome (unapproved): 500 mg PO tid. Glucophage XR: 500 mg PO daily with evening meal; increase by 500 mg q week to max 2000 mg/day (may divide 1000 mg PO bid). Fortamet: 500-1000 mg PO daily with evening meal; increase by 500 mg q week to max 2500 mg/day. [Generic/Trade: Tabs 500, 850,1000 mg, extended release 500, 750 mg. Trade only, extended release: Fortamet 500, 1000 mg; Glumetza 500, 1000 mg. Trade only: oral soln 500 mg/5 mL (Riomet).] ▶K ♀B ▶? $

pramlintide (*Symlin*): Type 1 DM with mealtime insulin therapy: Initiate 15 mcg SC immediately before major meals & titrate by 15 mcg increments (if significant nausea has not occurred for ≥3 days) to maintenance 30-60 mcg as tolerated. Type 2 DM with mealtime insulin therapy: Initiate 60 mcg SC immediately before major meals and increase to 120 mcg as tolerated (if significant nausea has not occurred for 3-7 days). [Trade only: 5 mL vials, 0.6 mg/mL.] ▶K ♀C ▶? $$$$

Gout-Related

allopurinol (**Aloprim, Zyloprim, ♣Purinol**): Mild gout or recurrent calcium oxalate stones: 200-300 mg PO daily-bid, max 800 mg/day. [Generic/Trade: Tabs 100, 300 mg.] ▶K ♀C ▶+ $

Colbenemid (colchicine + probenecid): 1 tab PO daily x 1 week, then 1 tab PO bid. [Generic/Trade: Tabs 0.5 mg colchicine + 500 mg probenecid.] ▶KL ♀C ▶? $$

colchicine: Rapid treatment of acute gouty arthritis: 0.6 mg PO q1h for up to 3h (max 3 tabs). Gout prophylaxis: 0.6 mg PO bid if CrCl ≥50 mL/min, 0.6 mg PO daily if CrCl 35-49 mL/min, 0.6 mg PO q2-3 days if CrCl 10-34 mL/min. [Generic/Trade: Tabs 0.6 mg.] ▶L ♀C ▶? $

probenecid (**♣Benuryl**): Gout: 250 mg PO bid x 7 days, then 500 bid. Adjunct to penicillin injection: 1-2 g PO. [Generic: Tabs 500 mg.] ▶KL ♀B ▶? $

Minerals

calcium acetate (**PhosLo**): Hyperphosphatemia: Initially 2 tabs/caps PO with each meal. [Trade only: Tab/Cap 667 mg (169 mg elem Ca).] ▶K ♀+ ▶? $

calcium carbonate (**Tums, Os-Cal, Caltrate, Viactiv, ♣Calsan**): 1-2 g elem Ca/day or more PO with meals divided bid-qid. [OTC Generic/Trade: tab 650,667,1250,1500 mg, chew tab 750,1250 mg, cap 1250 mg, susp 1250 mg/5 mL. Calcium carbonate is 40% elem Ca and contains 20 mEq of elem Ca/g calcium carbonate. Not more than 500-600 mg elem Ca/dose. Os-Cal 250 + D contains 125 units vitamin D/tab, Os-Cal 500 + D contains 200 units vitamin D/tab, Caltrate 600 + D contains 200 units vitamin D/tab. Viactiv (chew candy) 500 +100 units vitamin D + 40 mcg vitamin K/tab.] ▶K ♀+ ▶+ $

calcium chloride: 500-1000 mg slow IV q1-3 days. [Generic: injectable 10% (1000 mg/10 mL) 10 mL ampules, vials, syringes.] ▶K ♀+ ▶+ $

calcium citrate (**Citracal**): 1-2 g elem Ca/day or more PO with meals divided bid-qid. [OTC: Trade only (elem Ca): tab 200 mg, 250 mg with 125 units vitamin D, 315 mg with 200 units vitamin D, 250 mg with 62.5 units magnesium stearate vitamin D, effervescent tab 500 mg.] ▶K ♀+ ▶+ $

calcium gluconate: 2.25-14 mEq slow IV. 500-2000 mg PO bid-qid. [Generic: injectable 10% (1000 mg/10 mL, 4.65mEq/10 mL) 10,50,100,200 mL. OTC generic: tab 500,650,975,1000 mg.] ▶K ♀+ ▶+ $

ferrous gluconate (**Fergon**): 800-1600 mg ferrous gluconate PO divided tid. [OTC Generic/Trade: tab (ferrous gluconate) 240,300,324,325 mg.] ▶K ♀+ ▶+ $

ferrous sulfate (**Fer-in-Sol, FeoSol Tabs, ♣Ferodan, Slow-Fe, Fero-Grad**): 500-1000 mg ferrous sulfate (100-200 mg elem iron) PO divided tid. Liquid: Adults 5-10 mL tid, non-infant children 2.5-5 mL tid. Many other available formulations. [OTC Generic/Trade: tab (mg ferrous sulfate): tab 324,325 mg, liquid 220 mg/5 mL, drops 75 mg/0.6 mL.] ▶K ♀+ ▶+ $

INTRAVENOUS SOLUTIONS	(ions in mEq/l)							
Solution	Dextrose	Cal/l	Na	K	Ca	Cl	Lactate	Osm
0.9 NS	0 g/l	0	154	0	0	154	0	310
LR	0 g/l	9	130	4	3	109	28	273
D5 W	50 g/l	170	0	0	0	0	0	253
D5 0.2 NS	50 g/l	170	34	0	0	34	0	320
D5 0.45 NS	50 g/l	170	77	0	0	77	0	405
D5 0.9 NS	50 g/l	170	154	0	0	154	0	560
D5 LR	50 g/l	179	130	4	3	109	28	527

fluoride (*Luride*, ❖*Fluor-A-Day*, *Fluotic*): Adult dose: 10 mL of topical rinse swish and spit daily. Peds daily dose based on fluoride content of drinking water (table). [Generic: chew tab 0.5,1 mg, tab 1 mg, drops 0.125 mg, 0.25 mg, and 0.5 mg/ dropperful, lozenges 1 mg, solution 0.2 mg/mL, gel 0.1%, 0.5%, 1.23%, rinse (sodium fluoride) 0.05,0.1,0.2%).] ▶K ♀? ▷? $

Age	<0.3	0.3-0.6	>0.6
(years)	ppm	ppm	ppm
0-0.5	none	none	none
0.5-3	0.25mg	none	none
3-6	0.5 mg	0.25mg	none
6-16	1 mg	0.5 mg	none

Peds daily dose is based on drinking water fluoride (shown in ppm)

iron dextran (*InFed*, *DexFerrum*, ❖*Dexiron*, *Infufer*): 25-100 mg IM daily prn. Equations available to calculate IV dose based on weight & Hb. ▶KL ♀- ▷? $$$$

iron polysaccharide (*Niferex*, *Niferex-150*, *Nu-Iron*, *Nu-Iron 150*): 50-200 mg PO divided daily-tid. [OTC Generic: cap 150 mg (Niferex). Generic/Trade: cap 150 mg (Niferex-150, Nu-Iron 150), liquid 100 mg/5 mL (Niferex, Nu-Iron). 1 mg iron polysaccharide = 1 mg elemental iron.] ▶K ♀+ ▷+ $

iron sucrose (*Venofer*): Iron deficiency with hemodialysis: 5 mL (100 mg elem iron) IV over 5 min or diluted in 100 mL NS IV over ≥15 min. Iron deficiency in non-dialysis chronic kidney disease: 10 mL (200 mg elem iron) IV over 5 minutes. ▶KL ♀B ▷? $$$$$

magnesium chloride (*Slow-Mag*): 2 tabs PO daily. [OTC Trade only: enteric coated tab 64 mg. 64 mg tab Slow-Mag = 64 mg elem magnesium.] ▶K ♀A ▷+ $

magnesium gluconate (*Almora*, *Magtrate*, *Maganate*, ❖*Maglucate*): 500-1000 mg PO divided tid. [OTC Generic: tab 500 mg, liquid 54 mg elem Mg/5 mL.] ▶K♀A▷+$

magnesium oxide (*Mag-200*, *Mag-Ox 400*): 400-800 mg PO daily. [OTC Generic/Trade: cap 140,250,400,420,500 mg.] ▶K ♀A ▷+ $

magnesium sulfate: Hypomagnesemia: 1 g of 20% soln IM q6h x 4 doses, or 2 g IV over 1 h (monitor for hypotension). Peds: 25-50 mg/kg IV/IM q4-6h for 3-4 doses, max single dose 2g. ▶K ♀A ▷+ $

phosphorus (*Neutra-Phos*, *K-Phos*): 1 cap/packet PO qid. 1-2 tabs PO qid. Severe hypophosphatemia (eg, <1 mg/dl): 0.08-0.16 mmol/kg IV over 6h. [OTC: Trade only: (Neutra-Phos, Neutra-Phos K) tab/cap/packet 250 mg (8 mmol) phosphorus. Rx: Trade only: (K-Phos) tab 250 mg (8 mmol) phosphorus.] ▶K ♀C ▷? $

potassium (*Cena-K*, *Effer-K*, *K-8*, *K+10*, *Kaochlor*, *Kaon*, *Kaon Cl*, *Kay Ciel*, *Kaylixir*, *K+Care*, *K+Care ET*, *K-Dur*, *K-G Elixir*, *K-Lease*, *K-Lor*, *Klor-con*, *Klorvess*, *Klorvess Effervescent*, *Klotrix*, *K-Lyte*, *K-Lyte Cl*, *K-Norm*, *Kolyum*, *K-Tab*, *K-vescent*, *Micro-K*, *Micro-K LS*, *Slow-K*, *Ten-K*, *Tri-K*): IV infusion 10 mEq/h (diluted). 20-40 mEq PO daily-bid. [See table below for various forms; different salts include chloride, bicarbonate, citrate, acetate, gluconate. Potassium gluconate is available OTC.] ▶K ♀C ▷? $

POTASSIUM, oral forms

Effervescent Granules: 20 mEq (Klorvess Effervescent, K-vescent)
Effervescent Tablets: 20 mEq (K+Care ET, Klorvess), 25 mEq (Effer-K, K+Care ET, K-Lyte, K-Lyte/Cl, Klor-Con/EF), 50 mEq (K-Lyte DS, K-Lyte/Cl 50)
Liquids: 20 mEq/15 ml (Cena-K, Kaochlor S-F, K-G Elixir, Kaochlor 10%, Kay Ciel, Kaon, Kaylixir, Klorvess, Kolyum, Potasalan, Twin-K), 30 mEq/15 ml (Rum-K), 40 mEq/15 ml (Cena-K, Kaon-Cl 20%), 45 mEq/15 ml (Tri-K)
Powders: 15 mEq/pack (K+Care), 20 mEq/pack (Gen-K, K+Care, Kay Ciel, K-Lor, Klor-Con), 25 mEq/pack (K+Care, Klor-Con 25)
Tabs/Caps: 8 mEq (K+8, Klor-Con 8, Micro-K), 10 mEq (K+10, K-Norm, Kaon-Cl 10, Klor-Con 10, Klotrix, K-Tab, K-Dur 10, Micro-K 10), 20 mEq (Klor-Con M20, K-Dur20)

sodium ferric gluconate complex (**Ferrlecit**): 125 mg elem iron IV over 10 min or diluted in 100 mL NS IV over 1 h. Peds ≥6 yo: 1.5 mg/kg (max 125 mg) elem iron diluted in 25 mL NS & administered IV over 1h. ▶KL ♀B ▶? \$\$\$\$\$

PEDIATRIC REHYDRATION SOLUTIONS (ions in mEq/l)									
Brand	*Glucose*	*Cal/l*	*Na*	*K*	*Cl*	*Citrate*	*Phos*	*Ca*	*Mg*
CeraLyte 50*	0 g/l	160	50	20	40	30	0	0	0
CeraLyte 70*	0 g/l	160	70	20	60	30	0	0	0
CeraLyte 90*	0 g/l	160	90	20	80	30	0	0	0
Infalyte	30 g/l	140	50	25	45	34	0	0	0
Kao Lectrolyte*	20 g/l	90	50	20	40	30	0	0	0
Lytren†	20 g/l	80	50	25	45	30	0	0	0
Naturalyte	25 g/l	100	45	20	35	48	0	0	0
Pedialyte‡	25 g/l	100	45	20	35	30	0	0	0
Rehydralyte	25 g/l	100	75	20	65	30	0	0	0
Resol	20 g/l	80	50	20	50	34	5	4	4

*Available in premeasured powder packet. †Canada. ‡and Pedialyte Freezer Pops

Nutritionals

banana bag, rally pack: Alcoholic malnutrition (one formula): Add thiamine 100 mg + folic acid 1 mg + IV multivitamins to 1 liter NS and infuse over 4h. Magnesium sulfate 2g may be added. "Banana bag" and "rally pack" are jargon and not valid drug orders; specify individual components. ▶KL ♀+ ▶+ \$

fat emulsion (**Intralipid, Liposyn**): dosage varies. ▶L ♀C ▶? \$\$\$\$\$

formulas - infant (**Enfamil, Similac, Isomil, Nursoy, Prosobee, Soyalac, Alsoy, Nutramigen Lipil**): Milk meals. [OTC: Milk-based (Enfamil, Similac, SMA) or soy-based (Isomil, Nursoy, ProSobee, Soyalac, Alsoy).] ▶L ♀+ ▶+ \$

levocarnitine (**Carnitor**): 10-20 mg/kg IV at each dialysis session. ▶KL♀B ▶? \$\$\$\$\$

omega-3 fatty acid (**fish oil, Promega, Cardio-Omega 3, Sea-Omega, Marine Lipid Concentrate, MAX EPA, Omacor, SuperEPA 1200**): Hypertriglyceridemia: 2-4 g EPA+DHA content daily; Omacor: 4 capsules PO daily or divided bid. Marine Lipid Concentrate, Super EPA 1200 mg cap contains EPA 360 mg + DHA 240 mg, daily dose = 4-8 caps. Omacor is only FDA approved fish oil. [Generic/Trade: cap, shown as EPA+DHA mg content, 240 (Promega Pearls), 300 (Cardi-Omega 3, Max EPA), 320 (Sea-Omega), 400 (Sea-Omega), 500 (Sea-Omega), 600 (Marine Lipid Concentrate, SuperEPA 1200), 740 mg (Omacor), 875 mg (SuperEPA 2000).] ▶L ♀? ▶? \$\$

Thyroid Agents

levothyroxine (**L-Thyroxine, Levolet, Levo-T, Levothroid, Levoxyl, Novothyrox, Synthroid, Thyro-Tabs, Unithroid, T4, ✚Eltroxin**): Start 100-200 mcg PO daily (healthy adults) or 12.5-50 mcg PO daily (elderly or CV disease), increase by 12.5-25 mcg/day at 3-8 week intervals. Usual maintenance dose 100-200 mcg/day, max 300 mcg/d. [Generic/Trade: Tabs 25, 50, 75, 88, 100, 112, 125, 150, 175, 200, 300 mcg. Trade only: Tabs 137mcg.] ▶L ♀A ▶+ \$

liothyronine (**T3, Cytomel, Triostat**): Start 25 mcg PO daily, max 100 mcg/day. [Trade only: Tabs 5, 25, 50 mcg.] ▶L ♀A ▶? \$\$

methimazole (**Tapazole**): Start 5-20 mg PO tid or 10-30 mg PO daily, then adjust. [Generic/Trade: Tabs 5, 10. Generic only: Tabs 15, 20 mg.] ▶L ♀D ▶+ \$\$\$

propylthiouracil (**PTU, ✚Propyl Thyracil**): Start 100 mg PO tid, then adjust. Thyroid storm: 200-300 mg PO qid, then adjust. [Generic: Tabs 50 mg.] ▶L ♀D (but preferred over methimazole) ▶+ $

sodium iodide I-131 (*Iodotope, Sodium Iodide I-131 Therapeutic*): Specialized dosing for hyperthyroidism and thyroid carcinoma. [Generic/Trade: Capsules & oral solution: radioactivity range varies at the time of calibration.] ▶K ♀X ▶- $$$$$

Vitamins

ascorbic acid (**vitamin C**, ✱*Redoxon*): 70-1000 mg PO daily. [OTC: Generic: tab 25,50,100,250,500,1000 mg, chew tab 100,250,500 mg, time-released tab 500 mg, 1000,1500 mg, time-released cap 500 mg, lozenge 60 mg, liquid 35 mg/0.6 mL, oral solution 100 mg/mL, syrup 500 mg/5 mL.] ▶K ♀C ▶+ $

calcitriol (*Rocaltrol, Calcijex*): 0.25-2 mcg PO daily. [Generic/Trade: Cap 0.25, 0.5 mcg. Oral soln 1 mcg/mL. Injection 1,2 mcg/mL.] ▶L ♀C ▶? $$$

cyanocobalamin (**vitamin B12, *Nascobal, B12***): Deficiency states: 100-200 mcg IM q month or 1000-2000 mcg PO daily for 1-2 weeks followed by 1000 mcg PO daily or 500 mcg intranasal weekly. [OTC Generic: tab 100,500,1000,5000 mcg; lozenges 100,250,500 mcg. Rx Trade only: nasal gel 500 mcg/0.1 mL, nasal spray 500 mcg/spray (2.3 mL).] ▶K ♀C ▶+ $

Diatx (folic acid + niacinamide + cobalamin + pantothenic acid + pyridoxine + d-biotin + thiamine + vitamin C + riboflavin): 1 tab PO daily. [Trade only: Each tab contains folic acid 5 mg + niacinamide 20 mg + cobalamin 1 mg + pantothenic acid 10 mg + pyridoxine 50 mg + d-biotin 300 mcg + thiamine 1.5 mg + vitamin C 60 mg + riboflavin 1.5 mg. Diatx Fe: adds 100 mg ferrous fumarate per tab. Diatx Zn adds 25 mg of zinc oxide per tab.] ▶LK ♀? ▶? $$$

dihydrotachysterol (**vitamin D, *DHT***): 0.2-1.75 mg PO daily. [Generic: tab 0.125,0.2, 0.4 mg, cap 0.125 mg, oral solution 0.2 mg/mL.] ▶L ♀C ▶? $$

doxercalciferol (*Hectorol*): Secondary hyperparathyroidism on dialysis: Oral: 10 mcg PO 3x/ week. May increase q8 weeks by 2.5 mcg/dose; max 60 mcg/week. IV: 4 mcg IV 3x/ week. May increase dose q8 weeks by 1-2 mcg/dose; max 18 mcg/week. Secondary hyperparathyroidism not on dialysis: Start 1 mcg PO daily, may increase by 0.5 mcg/dose q 2 weeks. Max 3.5 mcg/day. [Trade only: Caps 0.5, 2.5 mcg.] ▶L ♀B ▶? $$$$$

Folgard (folic acid + cyanocobalamin + pyridoxine): 1 tab PO daily. [Trade: folic acid 0.8 mg + cyanocobalamin 0.115 mg + pyridoxine 10 mg tab.] ▶K ♀? ▶? $

folic acid (*folate, Folvite*): 0.4-1 mg IV/IM/PO/SC daily. [OTC Generic: Tab 0.4,0.8 mg. Rx Generic 1 mg.] ▶K ♀A ▶+ $

Foltx (folic acid + cyanocobalamin + pyridoxine): 1 tab PO daily. [Trade only: folic acid 2.5 mg/ cyanocobalamin 2 mg/ pyridoxine 25 mg tab.] ▶K ♀A ▶+ $

multivitamins (*MVI*): OTC & Rx: Many different brands and forms available with and without iron (tab, cap, chew tab, drops, liquid). ▶LK ♀+ ▶+ $

Nephrocap (vitamin C + folic acid + niacin + thiamine + riboflavin + pyridoxine + pantothenic acid + biotin + cyanocobalamin): 1 cap PO daily. If on dialysis, take after treatment. [Generic/Trade: vitamin C 100 mg/folic acid 1 mg/ niacin 20 mg/ thiamine 1.5 mg/ riboflavin 1.7 mg/ pyridoxine 10 mg/ pantothenic acid 5 mg/ biotin 150 mcg/ cyanocobalamin 6 mcg] ▶K ♀? ▶? $

Nephrovite (vitamin C + folic acid + niacin + thiamine + riboflavin + pyridoxine + pantothenic acid + biotin + cyanocobalamin): 1 tab PO daily. If on dialysis, take after treatment. [Generic/Trade: vitamin C 60 mg/folic acid 1 mg/ niacin 20 mg/ thiamine 1.5 mg/ riboflavin 1.7 mg/ pyridoxine 10 mg/ pantothenic acid 10 mg/ biotin 300 mcg/ cyanocobalamin 6 mcg] ▶K ♀? ▶? $

niacin (**vitamin B3, Niacor, Slo-Niacin, Niaspan**): 10-500 mg PO daily. See cardiovascular section for lipid-lowering dose. [OTC: Generic: tab 50,100,250,500 mg, timed-release cap 125,250,400 mg, timed-release tab 250,500 mg, liquid 50 mg/5 mL. Trade only: 250,500,750 mg (Slo-Niacin). Rx: Trade only: tab 500 mg (Niacor), timed-release cap 500 mg, timed-release tab 500,750,1000 mg (Niaspan, $$).] ▶K ♀C ▶? $

paricalcitol (**Zemplar**): Prevention/treatment of secondary hyperparathyroidism with renal insufficiency: 1-2 mcg PO daily or 2-4 mcg PO 3 times/wk; increase dose by 1 mcg/day or 2 mcg/week until desired PTH level is achieved. Prevention/treatment of secondary hyperparathyroidism with renal failure (CrCl<15 mL/min): 0.04-0.1 mcg/kg (2.8-7 mcg) IV 3 times/wk at dialysis; increase dose by 2-4 mcg q 2-4 weeks until desired PTH level is achieved. Max dose 0.24 mcg/kg (16.8 mcg). [Trade only: Caps 1, 2, 4 mcg.] ▶L ♀C ▶? $$$$$

phytonadione (**vitamin K, Mephyton, AquaMephyton**): Single dose of 0.5-1 mg IM within 1h after birth. Excessive oral anticoagulation: Dose varies based on INR. INR 5-9: 1-2.5 mg PO (≤5 mg PO may be given if rapid reversal necessary); INR >9 with no bleeding: 5-10 mg PO; Serious bleeding & elevated INR: 10 mg slow IV infusion. Adequate daily intake 120 mcg (males) and 90 mcg (females). [Trade only: Tab 5 mg.] ▶L ♀C ▶+ $

pyridoxine (**vitamin B6, B6**): 10-200 mg PO daily. INH overdose: 1 g IV/IM q 30 min, total dose of 1 g for each gram of INH ingested. [OTC Generic: Tab 25,50,100 mg, timed-release tab 100 mg.] ▶K ♀A ▶+ $

riboflavin (**vitamin B2**): 5-25 mg PO daily. [OTC Generic: tab 25,50,100 mg.] ▶K ♀A ▶+ $

thiamine (**vitamin B1**): 10-100 mg IV/IM/PO daily. [OTC Generic: tab 50,100,250, 500 mg, enteric coated tab 20 mg.] ▶K ♀A ▶+ $

tocopherol (**vitamin E, ♣Aquasol E**): RDA is 22 units (natural, d-alpha-tocopherol) or 33 units (synthetic, d,l-alpha-tocopherol) or 15 mg (alpha-tocopherol). Max recommended 1000 mg (alpha-tocopherol). Antioxidant: 400-800 units PO daily. [OTC Generic: tab 200,400 units, cap 73.5, 100, 147, 165, 200, 330, 400, 500, 600, 1000 units, drops 50 mg/mL.] ▶L ♀A ▶? $

vitamin A: RDA: 900 mcg RE (retinol equivalents) (males), 700 mcg RE (females). Treatment of deficiency states: 100,000 units IM daily x 3 days, then 50,000 units IM daily for 2 weeks. 1 RE = 1 mcg retinol or 6 mcg beta-carotene. Max recommended daily dose 3000 mcg. [OTC: Generic: cap 10,000, 15,000 units. Trade only: tab 5,000 units. Rx: Generic: 25,000 units. Trade only: soln 50,000 units/mL.] ▶L ♀A (C if exceed RDA, X in high doses) ▶+ $

vitamin D (**vitamin D2, ergocalciferol, Calciferol, Drisdol, ♣Osteoforte**): Familial hypophosphatemia (Vitamin D Resistant Rickets): 12,000-500,000 units PO daily. Hypoparathyroidism: 50,000-200,000 units PO daily. Adequate daily intake adults: 19-50 yo: 5 mcg (200 units) ergocalciferol; 51-70 yo: 10 mcg (400 units); >70 yo: 15 mcg (600 units). [OTC: Trade only: soln 8000 units/mL. Rx: Trade only: cap 50,000 units, inj 500,000 units/mL.] ▶L ♀A (C if exceed RDA) ▶+ $

Other

bromocriptine (**Parlodel**): Start 1.25-2.5 mg PO qhs with food. Hyperprolactinemia: Usual effective dose 2.5-15 mg/day, max 40 mg/day. Acromegaly: Usual effective dose 20-30 mg/day, max 100 mg/day. [Generic/Trade] Tabs 2.5 mg. Caps 5 mg.] ▶L ♀B ▶- $$$$

cabergoline (**Dostinex**): Hyperprolactinemia: 0.25-1 mg PO twice/wk. [Generic/Trade] Tabs 0.5 mg.] ▶L ♀B ▶- $$$$$

calcitonin (*Calcimar, Miacalcin, Fortical*, ♣*Caltine*): Skin test before using injectable product: 1 unit intradermally and observe for local reaction. Osteoporosis: 100 units SC/IM or 200 units (1 spray) intranasal daily (alternate nostrils). Paget's disease: 50-100 units SC/IM daily. Hypercalcemia: 4 units/kg SC/IM q12h. May increase after 2 days to max of 8 units/kg q6h. [Trade only: nasal spray 200 units/activation in 3.7 mL bottle (minimum of 30 doses/bottle).] ▶Plasma ♀C ▶? $$$$

cinacalcet (*Sensipar*): Treatment of secondary hyperparathyroidism in dialysis patients: 30 mg PO daily. May titrate q 2-4 weeks through sequential doses of 60, 90, 120 & 180 mg daily to target intact parathyroid hormone level of 150-300 pg/mL. Treatment of hypercalcemia in parathyroid carcinoma: 30 mg PO bid. May titrate q 2-4 weeks through sequential doses of 30 mg bid, 60 mg bid, 90 mg bid & 90 mg tid-qid as necessary to normalize serum calcium levels. [Trade only: tab 30, 60, 90 mg] ▶LK ♀C ▶? $$$$$

conivaptan (*Vaprisol*): Euvolemic hyponatremia: Loading dose of 20 mg IV over 30 minutes, then continuous infusion 20 mg per 24 hours for 1 to 3 days. Titrate to desired serum sodium. ▶LK ♀C ▶? $$$$$

cosyntropin (*Cortrosyn*, ♣*Synacthen*): Rapid screen for adrenocortical insufficiency: 0.25 mg (0.125 mg if <2 yo) IM/ IV over 2 min; measure serum cortisol before and 30-60 min after. ▶L ♀C ▶? $

demeclocycline (*Declomycin*): SIADH: 600-1200 mg/day PO given in 3-4 divided doses. [Trade only: Tabs 150, 300 mg] ▶K, feces ♀D ▶- $$$$$

desmopressin (*DDAVP, Stimate*, ♣*Minirin*): Diabetes insipidus: 10-40 mcg intranasally daily-tid, 0.05-1.2 mg/day PO or divided bid-tid, or 0.5-1 mL/day SC/IV in 2 divided doses. [Trade only: nasal solution 1.5 mg/mL (150 mcg/ spray). Generic/Trade: Tabs 0.1, 0.2 mg. Nasal solution 0.1 mg/mL (10 mcg/ spray). Note difference in concentration of nasal solutions.] ▶LK ♀B ▶? $$$$

growth hormone human (*Protropin, Genotropin, Norditropin, Norditropin NordiFlex, Nutropin, Nutropin AQ, Humatrope, Omnitrope, Serostim LQ, Saizen, Somatropin, Tev-Tropin, Zorbtive*): Dosages vary by indication. [Single dose vials (powder for injection with diluent). Tev-Tropin: 5mg vial (powder for injection with diluent, stable for 14 days when refrigerated). Genotropin: 1.5, 5.8, 13.8 mg cartridges. Humatrope: 6, 12, 24 mg per cartridges, 5mg vial (powder for injection with diluent, stable for 14 days when refrigerated). Nutropin AQ: 10 mg multiple dose vial & 10 mg/pen cartridges. Norditropin: 5,10,15 mg pen cartridges. Norditropin NordiFlex: 5, 10, 15 mg prefilled pens. Omnitrope: 1.5, 5.8 mg vial (powder for injection with diluent). Saizen: pre-assembled reconstitution device with autoinjector pen. Serostim LQ: 6 mg cartridge (packages come in 1 or 7 cartridges).] ▶LK ♀B/C ▶? $$$$$

hyaluronidase (*Amphadase*): Absorption and dispersion of injected drugs: Add 50-300 U (typically 150 U) to the injection solution. Hypodermoclysis: Administer 50-300 U after clysis or 150 U under the skin prior to clysis which will facilitate absorption of >1 L of solution. Subcutaneous urography: 75 U injected SC over each scapula, followed by injection of contrast media. [Trade only: 2 mL vial; 150 U/mL] ▶Serum ♀C ▶? $$

lanthanum carbonate (*Fosrenol*): Treatment of hyperphosphatemia in end stage renal disease: Start 750-1500 mg/day PO in divided doses with meals. Titrate dose every 2-3 weeks in increments of 750 mg/day until acceptable serum phosphate is reached. Most will require 1500-3000 mg/day to reduce serum phosphate <6.0 mg/dL. [Trade only: chewable tabs 250, 500, 750, 1000 mg.] ▶Not absorbed ♀C ▶? $$$$$

sevelamer (**Renagel**): Hyperphosphatemia: 800-1600 mg PO tid with meals. [Trade only: Tabs 400, 800 mg.] ▶Not absorbed ♀C ▶? $$$$$

sodium polystyrene sulfonate (**Kayexalate**): Hyperkalemia: 1 g/kg up to 15-60 g PO or 30-50 g retention enema (in sorbitol) q6h prn. Irrigate with tap water after enema to prevent necrosis. [Generic: Suspension 15 g/ 60 mL. Powdered resin.] ▶Fecal excretion ♀C ▶? $$$$

teriparatide (**Forteo**): Treatment of postmenopausal women or men with primary or hypogonadal osteoporosis and high risk for fracture: 20 mcg SC daily in thigh or abdomen for ≤2 years. [Trade only: 28-dose pen injector (20 mcg/dose).] ▶LK ♀C ▶- $$$$$

vasopressin (**Pitressin, ADH**, ✦**Pressyn AR**): Diabetes insipidus: 5-10 units IM/ SC bid-qid prn. Cardiac arrest: 40 units IV; may repeat if no response after 3 minutes. Septic shock: 0.01-0.1 units/min IV infusion, usual dose <0.04 units/min. Variceal bleeding: 0.2-0.4 units/min initially (max 0.9 units/min). ▶LK♀C ▶? $$$$$

ENT

Antihistamines - Nonsedating

desloratadine (**Clarinex**, ✦**Aerius**): Adults & children ≥12 yo: 5 mg PO daily. 6-11 yo: 1 teaspoonful (2.5 mg) PO daily. 12 mo - 5 yo: ½ teaspoonful (1.25 mg) PO daily. 6-11 mo: 2 mL (1 mg) PO daily. [Trade only: Tabs 5 mg. Fast-dissolve RediTabs 2.5 & 5 mg. Syrup 0.5 mg/mL] ▶LK ♀C ▶+ $$$

fexofenadine (**Allegra**): 60 mg PO bid or 180 mg daily. 6-12 yo: 30 mg PO bid. [Generic/Trade: Tabs 30, 60, 180 mg, Caps 60 mg.] ▶LK ♀C ▶+ $$$

loratadine (**Claritin, Claritin Hives Relief, Claritin RediTabs, Alavert, Tavist ND**): Adults & children ≥6 yo: 10 mg PO daily. 2-5 yo: 5 mg PO daily. [OTC: Generic/Trade: Tabs 10 mg. Fast-dissolve tabs (Alavert, Claritin RediTabs) 10 mg. Syrup 1 mg/mL.] ▶LK ♀B ▶+ $

Antihistamines - Other

azatadine (**Optimine**): 1-2 mg PO bid. [Trade: Tabs 1 mg, scored.] ▶LK ♀B ▶- $$$

cetirizine (**Zyrtec**, ✦**Reactine, Aller-Relief**): Adults & children ≥6 yo: 5-10 mg PO daily. 2-5 yo: 2.5 mg PO daily-bid. 6-23 mo: 2.5 mg PO daily. [Trade only: Tabs 5, 10 mg. Syrup 5 mg/5 mL. Chewable tabs, grape-flavored 5, 10 mg.] ▶LK ♀B ▶- $

chlorpheniramine (**Chlor-Trimeton, Chlo-Amine, Aller-Chlor**): 4 mg PO q4-6h. Max 24 mg/day. [OTC: Trade only: Chew tabs 2 mg. Generic/Trade: Tabs 4 mg. Syrup 2 mg/5mL. Tabs, Timed-release 8 mg. Trade only: Tabs, Timed-release 12 mg.] ▶LK ♀B ▶- $

clemastine (**Tavist**): 1.34 mg PO bid. Max 8.04 mg/day. [OTC: Generic/Trade: Tabs 1.34 mg. Rx: Generic/Trade: Tabs 2.68 mg. Trade only: Syrup 0.67 mg/5 mL.] ▶LK ♀B ▶- $

ENT COMBINATIONS abbreviations and footnotes

AC=acrivastine	DL=desloratadine	GU=guaifenesin	PR=promethazine
BR=brompheniramine	DM=dextromethorphan	HY=hydrocodone	PS=pseudoephedrine
CH=chlorpheniramine	DBR=dexbrompheniramine	LO=loratadine	PT=phenyltoloxamine
CO=codeine	DPH=diphenhydramine	PE=phenylephrine	PY=pyrilamine
CX=carbinoxamine	FE=fexofenadine	PH=pheniramine	TR=triprolidine

*5 mL/dose if 6-11 yo. 2.5 mL if 2-5yo. †1 mL/dose if 10-18 mo. ¾ mL if 7-9 mo. ½ mL if 4-6 mo. ¼ mL if 1-3 months old. ‡10 mL/dose if 6-11 yo. 5 mL if 2-5 yo. 2.5 mL if 13-23 mo. 1.25 mL if 4-12 mo. ¶Also contains acetaminophen

ENT COMBINATIONS (selected)	Decon- gestant	Antihis- tamine	Anti- tussive	Typical Adult Doses
OTC				
Actifed Cold & Allergy	PS	TR	-	1 tab or 10 mL q4-6h
Actifed Cold & Sinus¶	PS	CH	-	2 tabs q 6h
Allerfrim, Aprodine	PS	TR	-	1 tab or 10 mL q4-6h
Benadryl Allergy/Cold¶	PS	DPH	-	2 tabs q 6h
Benadryl-D Allergy/Sinus Tablets	PS	DPH	-	1 tab q 4-6 h
Claritin-D 12 hour, Alavert D-12	PS	LO	-	1 tab q12h
Claritin-D 24 hour	PS	LO	-	1 tab daily
Dimetapp Cold & Allergy Elixir	PS	BR	-	20mL q 4h
Dimetapp Multi-Symp Cold & Allergy¶	PE	CH	-	2 tabs q4h
Mucinex-DM Extended-Release	-	-	GU,DM	1-2 tab q12h
Robitussin CF	PS	-	GU, DM	10 mL q4h*
Robitussin DM, Mytussin DM	-	-	GU, DM	10 mL q4h*
Robitussin PE, Guaituss PE	PS	-	GU	10 mL q4h*
Drixoral Cold & Allergy	PS	DBR	-	1 tab q12h
Drixoral Cold & Flu¶	PS	DBR	-	2 tab q 12h
Triaminic Cold & Allergy	PS	CH	-	20 mL q4-6h‡
Triaminic Cough	PS	-	DM	20 mL q4h‡
Rx Only				
Allegra-D12- hour	PS	FE	-	1 tab q12h
Allegra-D24-hour	PS	FE	-	1 tab daily
Bromfenex	PS	BR	-	1 cap q12h
Clarinex-D24-hour	PS	DL	-	1 tab daily
Codeprex	-	CH	CO	10 mL q12h
Deconamine	PS	CH	-	1 tab or 10 mL tid-qid
Deconamine SR, Chlordrine SR	PS	CH	-	1 tab q12h
Deconsal II	PS	-	GU	1-2 tabs q12h
Dimetane-DX	PS	BR	DM	10 mL PO q4h
Duratuss	PS	-	GU	1 tab q12h
Duratuss HD©III	PS	-	GU, HY	10mL q4-6h
Entex PSE, Guaifenex PSE 120	PS	-	GU	1 tab q12h
Histussin D ©III	PS	-	HY	5 mL qid
Histussin HC ©III	PE	CH	HY	10 mL q4h
Humibid DM	-	-	GU, DM	1 tab q12h
Hycotuss©III	-	-	GU, HY	5mL pc & qhs
Phenergan/Dextromethorphan	-	PR	DM	5 mL q4-6h
Phenergan VC	PE	PR	-	5 mL q4-6h
Phenergan VC w/codeine©V	PE	PR	CO	5 mL q4-6h
Poly-Histine Elixir	-	PT/PY/PH	-	10 mL q4h*
Robitussin AC ©V	-	-	GU, CO	10 mL q4h*
Robitussin DAC ©V	PS	-	GU, CO	10 mL q4h*
Rondec Syrup	PS	BR	-	5 mL qid*
Rondec DM Syrup	PS	BR	DM	5 mL qid*
Rondec Oral Drops	PS	CX	-	0.25 to 1 mL qid†
Rondec DM Oral Drops	PS	CX	DM	0.25 to 1 mL qid†
Rynatan	PE	CH	-	1-2 tabs q12h
Rynatan-P Pediatric	PE	CH, PY	-	2.5-5 mL q12h*
Semprex-D	PS	AC	-	1cap q4-6h
Tanafed	PS	CH	-	10-20 mL q12h*
Triacin-C, Actifed w/codeine ©V	PS	TR	CO	10 mL q4-6h
Tussionex©III	-	CH	HY	5 mL q12h

cyproheptadine (**Periactin**): Start 4 mg PO tid. Max 32 mg/day. [Generic only: Tabs 4 mg. Syrup 2 mg/5 mL.] ▶LK ♀B ▶- $

dexchlorpheniramine (**Polaramine**): 2 mg PO q4-6h. Timed release tabs: 4 or 6 mg PO at qhs or q8-10h. [Trade only: Tabs, Immediate Release 2 mg. Syrup 2 mg/5 mL. Generic/Trade: Tabs, Timed Release: 4, 6 mg.] ▶LK ♀? ▶- $$$

diphenhydramine (**Benadryl, Dytan, ✲Allerdryl, Allernix, Nytol**): 25-50 mg IV/IM/PO q4-6h. Peds: 5 mg/kg/day divided q4-6h. [OTC: Trade only: Tabs 25, 50 mg, chew tabs 12.5 mg. OTC & Rx: Generic/Trade: Caps 25, 50 mg, softgel cap 25 mg. OTC: Generic/Trade: Liquid 6.25 or 12.5 mg per 5 mL. Rx only (Dytan): Suspension 25 mg/mL.] ▶LK ♀B ▶- $

hydroxyzine (**Atarax, Vistaril**): 25-100 mg IM/PO daily-qid or prn. [Generic/Trade: Tabs 10, 25, 50 mg. Trade only: 100 mg. Generic/Trade: Caps 25, 50, 100 mg. Syrup 10 mg/5 mL (Atarax). Trade only: Susp 25 mg/5 mL (Vistaril).] ▶L ♀C ▶- $

promethazine (**Phenergan**): Adults: 12.5-25 mg PO/IM/IV/PR daily-qid. Peds: 6.25-12.5 mg PO/IM/IV/PR daily-qid. [Trade only: Tabs 12.5 mg, scored. Generic/Trade: Tabs 25, 50 mg. Syrup 6.25 mg/5 mL. Trade: Phenergan Fortis Syrup 25 mg/5 mL. Suppositories 12.5 & 25 mg. Generic/Trade: Supp 50 mg].] ▶LK ♀C ▶- $

Antitussives / Expectorants

benzonatate (**Tessalon, Tessalon Perles**): 100-200 mg PO tid. Swallow whole. Do not chew. Numbs mouth; possible choking hazard. [Generic/Trade: Softgel caps: 100, 200 mg.] ▶L ♀C ▶? $

dextromethorphan (**Benylin, Delsym, Vick's**): 10-20 mg PO q4h or 30 mg PO q6-8h. Sustained action liquid 60 mg PO q12h. [OTC: Trade only: Caps 30 mg. Lozenges 2.5, 5, 7.5, 15 mg. Liquid 3.5, 5, 7.5, 10, 12.5, 15 mg/5 mL; 10 & 15 mg/15 mL (Generic/Trade). Trade only (Delsym): Sustained action liquid 30 mg/5 mL.] ▶L ♀+ ▶+ $

guaifenesin (**Robitussin, Hytuss, Guiatuss, Mucinex**): 100-400 mg PO q4h. 600-1200 mg PO q12h (extended release). 100-200 mg/dose if 6-11. 50-100 mg/dose if 2-5 yo. [OTC: Trade only: Tabs 100, 200 mg. Caps 200 mg. Extended release tabs 600 & 1200 mg. Syrup 100 & 200 mg/5 mL. Generic: Syrup 100 mg/5 mL. Rx: Trade only: Tabs 200, 1200 mg. Syrup 100 mg/5 mL.] ▶L ♀C ▶+ $

Decongestants (See ENT - Nasal Preparations for nasal spray decongestants)

phenylephrine (**Sudafed PE**): 10 mg PO q4h. [OTC trade tabs 10 mg.] ▶L ♀C ▶+ $

pseudoephedrine (**Sudafed, Sudafed 12 Hour, Efidac/24, Dimetapp Decongestant Infant Drops, PediaCare Infants' Decongestant Drops, Triaminic Oral Infant Drops, ✲Pseudofrin**): Adult: 60 mg PO q4-6h. Peds: 30 mg/dose if 6-12 yo, 15 mg/dose if 2-5 yo. Extended release tabs: 120 mg PO bid or 240 mg PO daily. Dimetapp, PediaCare, & Triaminic Infant Drops: (7.5 mg/0.8 mL): Give PO q4-6h. Max 4 doses/day. 2-3 yo: 1.6 mL. 12-23 mo: 1.2 mL. 4-11 mo: 0.8 mL. 0-3 mo: 0.4 mL. [OTC: Generic/Trade: Tabs 30, 60 mg. Chewable tabs 15 mg. Trade only: Tabs, extended release 120, 240 mg. Trade only: Liquid 15 mg/5 mL. Generic/Trade: Liquid 30 mg/5 mL. Infant drops 7.5 mg/0.8 mL.] ▶L ♀C ▶+ $

Ear Preparations

Auralgan (benzocaine + antipyrine): 2-4 drops in ear(s) tid-qid prn. [Generic/Trade: Otic soln 10 & 15 mL.] ▶Not absorbed ♀C ▶? $

carbamide peroxide (**Debrox, Murine Ear**): 5-10 drops in ear(s) bid x 4 days. [OTC: Trade only: Otic soln 6.5%, 15 or 30 mL bottle.] ▶Not absorbed ♀? ▶? $

Cipro HC Otic (ciprofloxacin + hydrocortisone): ≥1 yo to adult: 3 drops in ear(s) bid x 7 days. [Trade only: Otic suspension 10 mL.] ▶Not absorbed ♀C ▶- $$$

Ciprodex Otic (ciprofloxacin + dexamethasone): ≥6 mo to adult: 4 drops in ear(s) bid x 7 days. [Trade only: Otic suspension 5 & 7.5 mL.] ▶Not absorbed ♀C ▶- $$$

Cortisporin Otic (hydrocortisone + polymyxin + neomycin): 4 drops in ear(s) tid-qid up to 10 days of soln or suspension. Peds: 3 drops in ear(s) tid-qid up to 10 days. Caveats with perforated TMs or tympanostomy tubes: (1) Risk of neomycin ototoxicity, especially if use prolonged or repeated; (2) Use susp rather than acidic soln. [Generic/Trade: Otic soln or susp 7.5, 10 mL.] ▶Not absorbed ♀? ▶? $$

Cortisporin TC Otic (hydrocortisone + neomycin + thonzonium + colistin): 4-5 drops in ear(s) tid-qid up to 10 days. [Trade only: Otic suspension, 10 mL.] ▶Not absorbed ♀? ▶? $$$

docusate sodium (*Colace*): Cerumen removal: Instill 1 mL in affected ear. [Generic/Trade: Liquid 150 mg/15 mL.] ▶Not absorbed ♀+ ▶+ $

Domeboro Otic (acetic acid + aluminum acetate): 4-6 drops in ear(s) q2-3h. Peds: 2-3 drops in ear(s) q3-4h. [Generic/Trade: Otic soln 60 mL.] ▶Not absorbed♀?▶?$

fluocinolone - otic: 5 drops in affected ear(s) bid for 7-14 days. [Generic: Otic oil 0.01% 4 mL vials, 3 per pack.] ▶L ♀C ▶? $

ofloxacin - otic (*Floxin Otic*): >12 yo: 10 drops in ear(s) bid. 1-12 yo: 5 drops in ear(s) bid. [Generic/Trade: Otic soln 0.3% 5, 10 mL. Trade only: single-dispensing containers 0.25 mL (5 drops), 2 per foil pouch.] ▶Not absorbed ♀C ▶- $$

Pediotic (hydrocortisone + polymyxin + neomycin): 3-4 drops in ear(s) tid-qid up to 10 days. [Trade only: Otic suspension 7.5 mL.] ▶Not absorbed ♀? ▶? $$$

Swim-Ear (isopropyl alcohol + anhydrous glycerins): 4-5 drops in ears after swimming. [OTC: Trade only: Otic soln 30 mL.] ▶Not absorbed ♀? ▶? $

triethanolamine (*Cerumenex*): Fill ear canal x 15-30 mins to loosen cerumen, then flush. [Trade only: Otic soln 6 & 12 mL with dropper.] ▶Not absorbed ♀C ▶- $$

VoSol otic (acetic acid + propylene glycol): 5 drops in ear(s) tid-qid. Peds >3 yo: 3-4 drops in ear(s) tid-qid. VoSoL HC adds hydrocortisone 1%. [Generic/Trade: Otic soln 2%, 15 & 30 mL.] ▶Not absorbed ♀? ▶? $

Mouth & Lip Preparations

amlexanox (*Aphthasol, OraDisc A*): Aphthous ulcers: Apply ¼ inch paste or mucoadhesive patch to affected area qid after oral hygiene for up to 10 days. Up to 3 patches may be applied at one time. [Trade only: Oral paste 5%, 5 g tube. Mucoadhesive patch 2 mg, #20.] ▶LK ♀B ▶? $

cevimeline (*Evoxac*): Dry mouth due to Sjogren's syndrome: 30 mg PO tid. [Trade only: Caps 30 mg.] ▶L ♀C ▶- $$$

chlorhexidine gluconate (*Peridex, Periogard,* ♣*Denticare*): Rinse with 15 mL of undiluted soln for 30 seconds bid. Do not swallow. Spit after rinsing. [Generic/Trade: Oral rinse 0.12% 473-480 mL bottles.] ▶Fecal excretion ♀B ▶? $

clotrimazole (*Mycelex*): Oral troches dissolved slowly in mouth 5 times/day x 14 days. [Trade only: Oral troches 10 mg.] ▶L ♀C ▶? $$$$

Debacterol (sulfuric acid + sulfonated phenolics): Aphthous stomatitis, mucositis: Apply to dry ulcer. Rinse with water. [Trade only: 1 mL prefilled, single-use applicator.] ▶Not absorbed ♀C ▶+ $$

docosanol (*Abreva*): Herpes labialis (cold sores): At 1st sign of infection apply 5x/day until healed. [OTC: Trade only: 10% cream in 2 g tube.] ▶? ♀B ▶? $

doxycycline (*Periostat*): Periodontitis, adjunct to scaling & root planing: 20 mg PO bid. [Generic/Trade: Caps 20 mg.] ▶L ♀D ▶- $$$

Gelclair (maltodextrin + propylene glycol): Aphthous ulcers, mucositis, stomatitis: Rinse mouth with 1 packet tid or prn. [Trade only: 21 packets/box.] ▶Not absorbed ♀+ ▶+ $$$

lidocaine viscous (*Xylocaine*): Mouth or lip pain in adults only: 15-20 mL topically or swish & spit q3h. [Generic/Trade: soln 2%, 20 mL unit dose, 50, 100, 450 mL bottles.] ▶LK ♀B ▶+ $

magic mouthwash (*Benadryl + Mylanta + Carafate*): 5 mL PO swish & spit or swish & swallow tid before meals and prn. [Compounded suspension. A standard mixture is 30 mL diphenhydramine liquid (12.5 mg/5 mL)/60 mL Mylanta or Maalox/4 g Carafate.] ▶LK ♀B(- in 1st trimester) ▶- $$$

nystatin (*Mycostatin*, ✦*Nilstat*): Thrush: 5 mL PO swish & swallow qid. Infants: 2 mL/dose with 1 mL placed in each cheek. [Generic/Trade: Suspension 100,000 units/mL 60 & 480 mL. Trade only: Oral lozenges (Pastilles) 200,000 units.] ▶Not absorbed ♀C ▶? $$

penciclovir (*Denavir*): Herpes labialis (cold sores): apply cream q2h while awake x 4 days. Start at first sign of symptoms. [Trade only: Cream 1%, 1.5 g tube.] ▶Not absorbed ♀B ▶- $$

pilocarpine (*Salagen*): Dry mouth due to radiation of head & neck or Sjogren's syndrome: 5 mg PO tid-qid. [Generic/Trade: Tabs 5, 7.5 mg.] ▶L ♀C ▶- $$$

triamcinolone - topical (*Kenalog in Orabase*, ✦*Oracort*): Apply paste bid-tid, ideally pc & qhs. [Generic/Trade: 0.1% oral paste, 5 g tubes.] ▶L ♀C ▶? $

Nasal Preparations

azelastine (*Astelin*): Allergic/vasomotor rhinitis: 1-2 sprays/nostril bid. [Trade only: Nasal spray, 100 sprays/bottle.] ▶L ♀C ▶? $$$

beclomethasone (*Vancenase, Vancenase AQ Double Strength, Beconase AQ*): Vancenase: 1 spray per nostril bid-qid. Beconase AQ: 1-2 spray(s) per nostril bid. Vancenase AQ Double Strength: 1-2 spray(s) per nostril bid. [Trade only: Vancenase 42 mcg/spray, 80 or 200 sprays/bottle. Beconase AQ 42 mcg/spray, 200 sprays/bottle. Vancenase AQ Double Strength 84 mcg/spray, 120 sprays/bottle.] ▶L ♀C ▶? $$$

budesonide - nasal (*Rhinocort Aqua*): 1-4 sprays per nostril daily. [Trade only: Nasal inhaler 120 sprays/bottle.] ▶L ♀B ▶? $$$

cromolyn (*NasalCrom, BenaMist, Children's NasalCrom*): 1 spray per nostril tid-qid. [OTC: Generic/Trade: Nasal inhaler 200 sprays/bottle. Trade: Children's NasalCrom with special "child-friendly" applicator. 100 sprays/bottle.] ▶LK ♀B ▶+ $

flunisolide (*Nasalide, Nasarel*, ✦*Rhinalar*): Start 2 sprays/nostril bid. Max 8 sprays/nostril/day. [Generic/Trade: Nasal soln 0.025%, 200 sprays/bottle. Nasalide with pump unit. Nasarel with meter pump & nasal adapter.] ▶L ♀C ▶? $$

fluticasone (*Flonase*): 2 sprays per nostril daily. [Generic/Trade: Nasal spray 0.05%, 120 sprays/bottle.] ▶L ♀C ▶? $$$

ipratropium (*Atrovent Nasal Spray*): 2 sprays per nostril bid-qid. [Generic/Trade: Nasal spray 0.03%, 345 sprays/bottle & 0.06%, 165 sprays/bottle.] ▶L ♀B ▶? $$

levocabastine (✦*Livostin Nasal Spray*): Canada only. 2 sprays in each nostril bid, increase prn to tid-qid. [Trade only: nasal spray 0.5mg/mL, plastic bottles of 15ml. Each spray delivers 50 mcg.] ▶L (but minimal absorption) ♀C ▶- $$

mometasone - nasal (*Nasonex*): Adult: 2 sprays/nostril daily. Peds 2-11 yo: 1 spray/nostril daily. [Trade only: Nasal spray, 120 sprays/bottle.] ▶L ♀C ▶? $$$

oxymetazoline (*Afrin, Dristan 12 Hr Nasal, Nostrilla*): 2-3 drops/sprays per nostril bid prn rhinorrhea for ≤ 3 days. [OTC: Generic/Trade: Nasal spray 0.05%, 15 & 30 mL bottles. Nose drops 0.025% & 0.05%, 20 mL with dropper.] ▶L ♀C ▶? $

phenylephrine - nasal (*Neo-Synephrine, Sinex, Nostril*): 2-3 sprays/drops per nostril q4h prn x 3 days. [OTC: Trade only: Nasal drops 0.125, 0.16%. Generic/Trade: Nasal spray/drops 0.25, 0.5, 1%.] ▶L ♀C ▶? $

saline nasal spray (**SeaMist, Pretz, NaSal, Ocean, ♥HydraSense**): Nasal dryness: 1-3 sprays or drops per nostril prn. [Generic/Trade: Nasal spray 0.4, 0.5, 0.6, 0.65, 0.75%. Nasal drops 0.4, 0.65, & 0.75%.] ▶Not metabolized ♀A ▶+ $

triamcinolone - nasal (**Nasacort AQ, Nasacort HFA, Tri-Nasal**): Nasacort HFA, Tri-Nasal: 2 sprays per nostril daily-bid. Max 4 sprays/nostril/day. Nasacort AQ: 2 sprays per nostril daily. [Trade only: Nasal inhaler 55 mcg/spray, 100 sprays/bottle (Nasacort HFA). Nasal spray, 55 mcg/spray, 120 sprays/bottle (Nasacort AQ). Nasal spray 50 mcg/spray, 120 sprays/bottle (Tri-Nasal).] ▶L ♀C ▶- $$$

Other

Cetacaine (benzocaine + tetracaine + butyl aminobenzoate): Topical anesthesia of mucous membranes: Spray: Apply for ≤1 second. Liquid or gel: Apply with cotton applicator directly to site. [Trade only: Spray 56g. Liquid 56 g. Hospital gel 29 g.] ▶LK ♀C ▶? $

GASTROENTEROLOGY

Antidiarrheals

bismuth subsalicylate (**Pepto-Bismol, Kaopectate**): 2 tabs or 30 mL (262 mg/15 mL) PO q 30 min-1 h up to 8 doses/day. Peds: 10 mL (262 mg/15 mL) or 2/3 tab if 6-9 yo, 5 mL (262 mg/15 mL) or 1/3 tab if 3-6 yo. Risk of Reye's syndrome in children. [OTC Generic/Trade: chew tab 262 mg, susp 262,525,750 mg/15 mL. Generic: susp 130 mg/15 mL. Trade: caplets 262 mg (Pepto-Bismol), susp 87 mg/5 mL (Kaopectate Children's Liquid), caplets 750mg (Kaopectate).] ▶K ♀D ▶? $

Imodium Advanced (loperamide + simethicone): 2 caplets PO initially, then 1 caplet PO after each unformed stool to a max of 4 caplets/24 h. Peds: 1 caplet PO initially, then ½ caplet PO after each unformed stool to a max of 2 caplets/day (if 6-8 yo or 48-59 lbs) or 3 caplets/day (if 9-11 yo or 60-95 lbs). [OTC Trade only: caplet & chew tab 2 mg loperamide/125 mg simethicone.] ▶L ♀B ▶+ $

Lomotil (diphenoxylate + atropine): 2 tabs or 10 mL PO qid. [Generic/Trade: soln 2.5/0.025 mg diphenoxylate/atropine per 5 mL, tab 2.5/0.025 mg.] ▶L ♀C ▶- ©V $

loperamide (**Imodium, Imodium AD, ♥Loperacap, Diarr-eze**): 4 mg PO initially, then 2 mg PO after each unformed stool to a maximum of 16 mg/day. Peds: 2 mg PO tid if >30 kg, 2 mg bid if 20-30 kg, 1 mg tid if 13-20 kg. [Rx Generic/Trade: cap 2 mg, tab 2 mg. OTC Generic/Trade: tab 2 mg, liquid 1 mg/5 mL.] ▶L ♀B ▶+ $

Motofen (difenoxin + atropine): 2 tabs PO initially, then 1 after each loose stool q3-4 h prn. Maximum of 8 tabs/24h. [Trade only: tab difenoxin 1 mg + atropine 0.025 mg.] ▶L ♀C ▶- ©IV $

opium (**opium tincture, paregoric**): 5-10 mL paregoric PO daily-qid or 0.3-0.6 mL PO opium tincture qid. [Trade only: opium tincture 10% (deodorized opium tincture, 10 mg morphine equivalent per mL). Generic: paregoric (camphorated opium tincture, 2 mg morphine equivalent/5 mL).] ▶L ♀B (D with long-term use) ▶? ©II (opium tincture), III (paregoric) $$

Antiemetics - 5-HT3 Receptor Antagonists

dolasetron (**Anzemet**): Nausea with chemo: 1.8 mg/kg up to 100 mg IV/PO single dose. Post-op nausea: 12.5 mg IV in adults and 0.35 mg/kg IV in children as single dose. Alternative for prevention 100 mg (adults) or 1.2 mg/kg (children) PO 2 h before surgery. [Trade only: tab 50,100, mg.] ▶LK ♀B ▶? $$$

granisetron (**Kytril**): Nausea with chemo: 10 mcg/kg IV over 5 minutes, 30 minutes prior to chemo. Oral: 1 mg PO bid x 1 day only. Radiation-induced nausea and vomiting: 2 mg PO 1 hr before first irradiation fraction of each day. [Trade only: tab 1 mg, oral soln 2 mg/10 mL (30 mL).] ▶L ♀B ▶? $$$

ondansetron (Zofran): Nausea with chemo (≥6 month old): 32 mg IV over 15 min, or 0.15 mg/kg doses 30 min prior to chemo and repeated at 4 & 8 hrs after first dose. Oral dose if ≥12 yo: 8 mg PO and repeated 8 hrs later. If 4-11 yo: 4 mg PO 30 min prior to chemo and repeated at 4 & 8 hrs. Prevention of post-op nausea: 4 mg IV over 2-5 min or 4 mg IM or 16 mg PO 1 hr before anesthesia. If 1 month-12 yo: 0.1 mg/kg IV over 2-5 min x 1 if ≤40 kg; 4 mg IV over 2-5 min x 1 if >40 kg. Prevention of N/V associated with radiotherapy: 8 mg PO tid. [Trade only: tab 4,8,24 mg, orally disintegrating tab 4, 8 mg, solution 4 mg/5 mL.] ▶L ♀B ▶? $$$$$

palonosetron (Aloxi): Nausea with chemo: 0.25 mg IV over 30 seconds, 30 minutes prior to chemo. ▶L ♀B ▶? $$$$$

Antiemetics - Other

aprepitant (Emend): Prevention of nausea with moderately to highly emetogenic chemo, in combination with dexamethasone and ondansetron: 125 mg PO on day 1 (1 h prior to chemo), then 80 mg PO qam on days 2 & 3. Prevention of postoperative N/V: 40 mg PO within 3 hours prior to anesthesia. [Trade only: cap 40, 80, 125 mg.] ▶L ♀B ▶? $$$$$

Diclectin (doxylamine + pyridoxine): Canada only. 2 tabs PO qhs. May add 1 tab in am and 1 tab in afternoon, if needed. [Trade only: delayed-release tab doxylamine 10 mg + pyridoxine 10 mg.] ▶LK ♀A ▶? $

dimenhydrinate (Dramamine, ✦Gravol): 50-100 mg PO/IM/IV q4-6h prn. [OTC: Generic/Trade: chew tab 50 mg. Trade only: chew tab 50 mg. Generic only: solution 12.5 mg/5ml.] ▶LK ♀B ▶- $

doxylamine (Unisom Nighttime Sleep Aid, others): 12.5 mg PO bid; often used in combination with pyridoxine. [OTC Generic/Trade: tab 25 mg.] ▶LK ♀A ▶? $

dronabinol (Marinol): Nausea with chemo: 5 mg/m² PO 1-3 h before chemo then 5 mg/m²/dose q2-4h after chemo for 4-6 doses/day. Anorexia associated with AIDS: Initially 2.5 mg PO bid before lunch and dinner. [Trade only: cap 2.5, 5, 10 mg.] ▶L ♀C ▶- ©III $$$$$

droperidol (Inapsine): 0.625-2.5 mg IV or 2.5 mg IM. May cause fatal QT prolongation, even in patients with no risk factors. Monitor ECG before. ▶L ♀C ▶? $

meclizine (Antivert, Bonine, Medivert, Meclicot, Meni-D, ✦Bonamine): Motion sickness: 25-50 mg PO 1 hr before travel. May repeat q24h prn. [Rx/OTC/Generic/Trade: tabs 12.5, 25 mg. Chew tabs 25 mg. Rx/Trade: tabs 50 mg.] ▶L ♀B ▶? $

metoclopramide (Reglan, ✦Maxeran): 10 mg IV/IM q2-3h prn. 10-15 mg PO qid, 30 min before meals and qhs. [Generic/Trade: tabs 5,10, mg, Generic only: soln 5 mg/5 mL.] ▶K ♀B ▶? $

nabilone (Cesamet): 1 to 2 mg PO bid, 1 to 3 hours before chemotherapy. [Trade: cap 1 mg.] ▶L ♀C ▶- ©II $$$$

phosphorated carbohydrates (Emetrol): 15-30 mL PO q15 min prn, max 5 doses. Peds: 5-10 mL. [OTC Generic/Trade: Solution containing dextrose, fructose, and phosphoric acid.] ▶L ♀A ▶+ $

prochlorperazine (Compazine, ✦Stemetil): 5-10 mg IV over at least 2 min. 5-10 mg PO/IM tid-qid. 25 mg PR q12h. Sustained release: 15 mg PO qam or 10 mg PO q12h. Peds: 0.1 mg/kg/dose PO/PR tid-qid or 0.1-0.15 mg/kg/dose IM tid-qid. [Generic/Trade: tabs 5,10,25 mg, supp 25 mg. Trade: ext'd-release caps (Compazine Spansules) 10,15,30 mg, supp 2.5,5 mg, liquid 5 mg/5 mL.] ▶LK ♀C ▶? $

promethazine (***Phenergan***): Adults: 12.5-25 mg PO/IM/PR q4-6h. Peds: 0.25-1 mg/kg PO/IM/PR q4-6h. Contraindicated if <2 yo; caution in older children. IV use common but not approved. [Generic/Trade: tab/supp 12.5,25,50 mg. Generic only: syrup 6.25 mg/5ml] ▶LK ♀C ▶- $

scopolamine (***Transderm-Scop, Scopace***, ♥***Transderm-V***): Motion sickness: Apply 1 disc (1.5 mg) behind ear 4h prior to event; replace q3 days. Tablet: 0.4 to 0.8 mg PO 1 hour before travel and q8h prn. [Trade only: topical disc 1.5 mg/72h, box of 4. Oral tablet 0.4 mg.] ▶LK ♀C ▶+ $

thiethylperazine (***Torecan***): 10 mg PO/IM 1-3 times/day. [Trade only: tab 10 mg.] ▶L ♀? ▶? $

trimethobenzamide (***Tigan***): 250 mg PO q6-8h, 200 mg IM/PR q6-8h. Peds: 100-200 mg/dose PO/PR q6-8h if 13.6-40.9 kg; 100 mg PR q6-8h if <13.6kg (not newborns). [Generic/Trade: Caps 300 mg.] ▶LK ♀C but + ▶? $

Antiulcer - Antacids

Alka-Seltzer (aspirin + citrate + bicarbonate): 2 regular strength tabs in 4 oz water q4h PO prn, max 8 tab (<60 yo) or 4 tabs (>60 yo) in 24h or 2 extra-strength tabs in 4 oz water q6h PO prn, max 7 tabs (<60 yo) or 4 tabs (>60 yo) in 24h. [OTC Trade only: regular strength, original: ASA 325 mg + citric acid 1000 mg + sodium bicarbonate 1916 mg. Regular strength lemon lime and cherry: 325 mg + 1000 mg + 1700 mg. Extra-strength: 500 mg + 1000 mg + 1985 mg. Not all forms of Alka Seltzer contain aspirin (eg, Alka Seltzer Heartburn Relief).] ▶LK ♀? (- 3rd trimester) ▶? $

aluminum hydroxide (***Alternagel, Amphojel, Alu-Tab, Alu-Cap***, ♥***Basalgel, Mucaine***): 5-10 mL or 1-2 tabs PO up to 6 times daily. Constipating. [OTC Generic/Trade: cap 475 mg, susp 320 & 600 mg/5 mL. Trade only: cap 400 mg (Alu-Cap)] ▶K ♀+ (? 1st trimester) ▶? $

calcium carbonate (***Tums, Mylanta Children's, Titralac, Rolaids Calcium Rich, Surpass***): 1000-3000 mg PO q2h prn or 1-2 pieces gum chewed prn, max 7000 mg/day. [OTC Generic/Trade: tabs 500, 600, 650, 750, 1000 mg, susp 1250 mg/5 mL, gum (Surpass) 300, 450 mg. Trade only: softchew 1177 mg (Rolaids). Generic only: sugar free tab 750 mg.] ▶K ♀+ (? 1st trimester) ▶? $

Citrocarbonate (bicarbonate + citrate): 1-2 teaspoonfuls in cold water PO 15 minutes to 2 hours after meals prn. [OTC Trade only: Na⁺ bicarbonate 0.78 g + Na⁺ citrate 1.82 g in each 1 teaspoonful dissolved in water, 150, 300g.] ▶K ♀? ▶? $

Gaviscon (aluminum hydroxide + magnesium carbonate): 2-4 tabs or 15-30 mL (regular strength) or 10 mL (extra strength) PO qid prn. [OTC Trade only: Tab: regular strength (Al hydroxide 80 mg + Mg trisilicate 20 mg), extra strength (Al hydroxide 160 mg + Mg carbonate 105 mg). Liquid: regular strength (Al hydroxide 95 mg + Mg carbonate 358 mg per 15 mL), extra strength (Al hydroxide 508 mg + Mg carbonate 475 mg per 30 mL.)] ▶K ♀? ▶? $

Maalox (aluminum hydroxide + magnesium hydroxide): 10-20 mL or 1-4 tab PO prn. [OTC Generic/Trade: regular strength chew tab (Al hydroxide + Mg hydroxide 200/200 mg), susp (225/200 mg per 5 mL).] ▶K ♀+ (? 1st trimester) ▶? $

magaldrate (***Riopan***): 5-10 mL PO prn. [OTC Trade only: susp 540 mg/5 mL. Riopan Plus (with simethicone) available as susp 540/20 mg/5 mL, chew tab 540/20 mg.] ▶K ♀+ (? 1st trimester) ▶? $

Mylanta (aluminum hydroxide + magnesium hydroxide + simethicone): 2-4 tabs or 10-45 mL PO prn. [OTC Generic/Trade: Liquid, double strength liquid, double strength tab. Trade: tab sodium + sugar + dye free.] ▶K ♀+ (? 1st trimester) ▶? $

Rolaids (calcium carbonate + magnesium hydroxide): 2-4 tabs PO q1h prn, max 12 tabs/day (regular strength) or 10 tabs/day (extra-strength). [OTC Trade only: Tab: regular strength (Ca carbonate 550 mg, Mg hydroxide 110 mg), extra-strength (Ca carbonate 675 mg, Mg hydroxide 135 mg).] ▶K ♀? ▶? $

Antiulcer - H2 Antagonists

cimetidine (***Tagamet, Tagamet HB***): 300 mg IV/IM/PO q6-8h, 400 mg PO bid, or 400-800 mg PO qhs. Erosive esophagitis: 800 mg PO bid or 400 mg PO qid. Continuous IV infusion 37.5-50 mg/h (900-1200 mg/day). [Rx Generic/Trade: tab 200, 300, 400, 800 mg. Rx Generic liquid 300 mg/5 mL. OTC Generic/Trade: tab 200 mg.] ▶LK ♀B ▶+ $$$

famotidine (***Pepcid, Pepcid AC, Maximum Strength Pepcid AC***): 20 mg IV q12h. 20-40 mg PO qhs, or 20 mg PO bid. [Generic/Trade: tab 10 mg (OTC, Pepcid AC Acid Controller), 20 (Rx and OTC, Maximum Strength Pepcid AC), 30, 40 mg. Rx Trade only: susp.40 mg/5 mL.] ▶LK ♀B ▶? $$

nizatidine (***Axid, Axid AR***): 150-300 mg PO qhs, or 150 mg PO bid. [OTC Trade only (Axid AR): tabs 75 mg Rx Trade only: oral solution 15 mg/mL (120, 480 mL). Rx Generic/Trade: cap 150, 300 mg.] ▶K ♀B ▶? $$$

Pepcid Complete (famotidine + calcium carbonate + magnesium hydroxide): Treatment of heartburn: 1 tab PO prn. Max 2 tabs/day. [OTC: Trade: chew tab famotidine 10 mg with Ca carbonate 800 mg & Mg hydroxide 165 mg.] ▶LK ♀B ▶? $

ranitidine (***Zantac, Zantac 25, Zantac 75, Zantac 150, Peptic Relief***): 150 mg PO bid or 300 mg PO qhs. 50 mg IV/IM q8h, or continuous infusion 6.25 mg/h (150 mg/d). [Generic/Trade: tabs 75 mg (OTC, Zantac 75, Zantac 150), 150,300 mg, syrup 75 mg/5 mL. Rx Trade only: effervescent tab 25,150 mg. Rx Generic only: caps 150,300 mg.] ▶K ♀B ▶? $$$

Antiulcer - Helicobacter pylori Treatment

Helidac (bismuth subsalicylate + metronidazole + tetracycline): 1 dose PO qid for 2 weeks. To be given with an H2 antagonist. [Trade only: Each dose: bismuth subsalicylate 524 (2x262 mg) chewable tab + metronidazole 250 mg tab + tetracycline 500 mg cap.] ▶LK ♀D ▶- $$$$

PrevPac (lansoprazole + amoxicillin + clarithromycin, ✿*HP-Pac*): 1 dose PO bid x 10-14 days. [Trade only: lansoprazole 30 mg x 2 + amoxicillin 1 g (2x500 mg) x 2, clarithromycin 500 mg x 2.] ▶LK ♀C ▶? $$$$$

HELICOBACTER PYLORI THERAPY

- Triple therapy PO x 7-14 days: clarithromycin 500 mg bid + amoxicillin 1 g bid (or metronidazole 500 mg bid) + a proton pump inhibitor*
- Quadruple therapy PO x 14 days: bismuth subsalicylate 525 mg (or 30 mL) tid-qid + metronidazole 500 mg tid-qid + tetracycline 500 mg tid-qid + a proton pump inhibitor* or a H₂ blocker†

*PPI's esomeprazole 40 mg qd, lansoprazole 30 mg bid, omeprazole 20 mg bid, pantoprazole 40 mg bid, rabeprazole 20 mg bid. †H₂ blockers cimetidine 400 mg bid, famotidine 20 mg bid, nizatidine 150 mg bid, ranitidine 150 mg bid. Adapted from *Medical Letter Treatment Guidelines* 2004.

Antiulcer - Proton Pump Inhibitors

esomeprazole (***Nexium***): Erosive esophagitis: 20-40 mg PO daily x 4-8 weeks. Maintenance of erosive esophagitis: 20 mg PO daily. GERD: 20 mg PO daily x 4 weeks. GERD with esophagitis: 20-40 mg IV daily x 10 days until taking PO. Prevention of NSAID-associated gastric ulcer: 20-40 mg PO daily x up to 6 months. H pylori eradication: 40 mg PO daily with amoxicillin 1000 mg PO bid & clarithromycin 500 mg PO bid x 10 days. [Trade only: delayed release cap 20, 40 mg.] ▶L ♀B ▶? $$$$

lansoprazole (**Prevacid, Prevacid NapraPac**): Duodenal ulcer or maintenance therapy after healing of duodenal ulcer, erosive esophagitis, NSAID-induced gastric ulcer: 30 mg PO daily x 8 weeks (treatment), 15 mg PO daily for up to 12 weeks (prevention). GERD: 15 mg PO daily. Gastric ulcer: 30 mg PO daily. Erosive esophagitis: 30 mg PO daily or 30 mg IV daily x 7 days or until taking PO. [Trade only: cap 15,30 mg. Susp 15,30 mg packets. Orally disintegrating tab 15,30 mg. Prevacid NapraPac: 7 lansoprazole 15 mg caps packaged with 14 naproxen tabs 375 mg or 500 mg.] ▶L ♀B ▶? $$$$

omeprazole (**Prilosec, Zegerid, Rapinex, ✦Losec**): Duodenal ulcer or erosive esophagitis: 20 mg PO daily. Heartburn (OTC): 20 mg PO daily x 14 days. Gastric ulcer: 40 mg PO daily. Hypersecretory conditions: 60 mg PO daily. [Rx Generic /Trade: Cap 10, 20 mg. OTC: 20 mg tab. Rx Trade only: Cap 40 mg, powder for suspension 20, 40 mg (Zegerid), oral suspension 20 mg (Rapinex).] ▶L ♀C ▶? $$$$

pantoprazole (**Protonix, ✦Pantoloc**): GERD: 40 mg PO daily. Zollinger-Ellison syndrome: 80 mg IV q8-12h x 6 days until taking PO. GERD associated with a history of erosive esophagitis: 40 mg IV daily x 7-10 days until taking PO. [Trade only: Tabs 20, 40 mg.] ▶L ♀B ▶? $$$

rabeprazole (**Aciphex, ✦Pariet**): 20 mg PO daily. [Trade only: Tab 20 mg.] ▶L ♀B ▶? $$$$

Antiulcer - Other

dicyclomine (**Bentyl, Bentylol, Antispas, ✦Formulex, Protylol, Lomine**): 10-20 mg PO/IM qid up to 40 mg PO qid. [Generic/Trade: Tab 20 mg, cap 10 mg, syrup 10 mg/5 mL. Generic only: cap 20 mg.] ▶LK ♀B ▶- $$

Donnatal (phenobarbital + atropine + hyoscyamine + scopolamine): 1-2 tabs/caps or 5-10 mL PO tid-qid. 1 extended release tab PO q8-12h. [Generic/Trade: Phenobarbital 16.2 mg + hyoscyamine 0.1 mg + atropine 0.02 mg + scopolamine 6.5 mcg in each tab, cap or 5 mL. Extended-release tab 48.6 + 0.3111 + 0.0582 + 0.0195 mg.] ▶LK ♀C ▶- $

"GI cocktail", "green goddess": Acute GI upset: mixture of Maalox/Mylanta 30 mL + viscous lidocaine (2%) 10 mL + Donnatal 10 mL PO in a single dose. ▶LK $

hyoscyamine (**Levsin, NuLev**): 0.125-0.25 mg PO/SL q4h prn. Sustained release: 0.375-0.75 mg PO q12h. [Generic/Trade: tab 0.125, SL tab 0.125 mg. solution 0.125 mg/mL, elixir 0.125 mg/5ml, extended-release tab/cap 0.375 mg, orally disintegrating tab 0.125 mg. Generic only: tab 0.15 mg.] ▶LK ♀C ▶- $

mepenzolate (**Cantil**): 25-50 mg PO qid, with meals and qhs. [Trade only: tab: 25 mg.] ▶LK ♀B ▶? $$$$$

misoprostol (**Cytotec**): NSAID-induced gastric ulcer prevention: Start 100 mcg PO bid, then titrate as tolerated up to 200 mcg qid. Abortifacient. Diarrhea in 13-40%, abdominal pain in 7-20%. [Generic/Trade: tab 100,200 mcg.] ▶LK ♀X ▶- $$$$

propantheline (**Pro-Banthine, ✦Propanthel**): 7.5-15 mg PO 30 min ac & qhs. [Generic/Trade: tab 15 mg. Trade only: tab 7.5 mg.] ▶LK ♀C ▶- $$$

simethicone (**Mylicon, Gas-X, Phazyme, ✦Ovol**): 40-160 mg PO qid prn. Infants: 20 mg PO qid prn [OTC: Generic/Trade: tab 60,95 mg, chew tab 40,80,125 mg, cap 125 mg, drops 40 mg/0.6 mL.] ▶Not absorbed ♀C but + ▶? $

sucralfate (**Carafate, ✦Sulcrate**): 1 g PO 1h before meals (2h before other medications) & qhs. [Generic/Trade: tab 1 g, susp 1g/10 mL.] ▶Not absorbed♀B▶? $$$

Laxatives - Bulk-Forming

methylcellulose (*Citrucel*): 1 heaping tablespoon in 8 oz. water PO daily-tid. [OTC Trade only: regular & sugar-free packets and multiple use canisters, clear-mix solution, caplets 500 mg.] ▶Not absorbed ♀+ ▶? $

polycarbophil (*FiberCon, Fiberall, Konsyl Fiber, Equalactin*): Laxative: 1 g PO qd prn. Diarrhea: 1 g PO q30 min. Max daily dose 6 g. [OTC Generic/Trade: tab 500,625 mg, chew tab 500,1000 mg.] ▶Not absorbed ♀+ ▶? $

psyllium (*Metamucil, Fiberall, Konsyl, Hydrocil, ✦Prodium Plain*): 1 tsp in liquid, 1 packet in liquid or 1-2 wafers with liquid PO daily-tid. [OTC: Generic/Trade: regular and sugar-free powder, granules, capsules, wafers, including various flavors and various amounts of psyllium.] ▶Not absorbed ♀+ ▶? $

Laxatives - Osmotic

glycerin (*Fleet*): one adult or infant suppository PR prn. [OTC Generic/Trade: supp infant & adult, solution (Fleet Babylax) 4 mL/applicator.] ▶Not absorbed ♀C ▶? $

lactulose (*Chronulac, Cephulac, Kristalose*): Constipation: 15-30 mL (syrup) or 10-20 g (powder for oral solution) PO daily. Hepatic encephalopathy: 30-45 mL (syrup) PO tid-qid, or 300 mL retention enema. [Generic/Trade: syrup 10 g/15 mL. Trade (Kristalose): 10, 20 g packets for oral solution.] ▶Not absorbed ♀B ▶? $$$

magnesium citrate (*✦Citro-Mag*): 150- 300 mL PO divided daily-bid. Children <6 yo: 2-4 mL/kg/24h. [OTC Generic: solution 300 mL/bottle. Low sodium & sugar-free available.] ▶K ♀+ ▶? $

magnesium hydroxide (*Milk of Magnesia*): Laxative: 30-60 mL regular strength liquid PO. Antacid: 5-15 mL regular strength liquid or 622-1244 mg PO qid prn. [OTC Generic/Trade: susp 400 mg/5mL. Chewable tab: chew tab 311 mg. Generic only: susp (concentrated) 1200 mg/5 mL, sugar-free 400 mg/5 mL.] ▶K ♀+ ▶? $

polyethylene glycol (*MiraLax, GlycoLax*): 17 g (1 heaping tablespoon) in 4-8 oz water, juice, soda, coffee, or tea PO daily. [Generic/Trade: powder for oral solution 17g/scoop. Trade: 17 g packets for oral solution.] ▶Not absorbed ♀C ▶? $

polyethylene glycol with electrolytes (*GoLytely, Colyte, TriLyte, NuLytely, Half-Lytely and Bisacodyl Tablet Kit, ✦Klean-Prep, Electropeg, Peg-Lyte*): Bowel prep: 240 mL q10 min PO until 4L is consumed. [Generic/Trade: powder for oral solution in disposable jug 4L. Also, as a kit of 2L bottle of polyethylene glycol with electrolytes and 4 bisacodyl tabs 5 mg (HalfLytely and Bisacodyl Tablet Kit). Trade only (GoLytely): packet for oral solution to make 3.785 L.] ▶Not absorbed ♀C ▶? $

sodium phosphate (*Fleet enema, Fleet Phospho-Soda, Accu-Prep, Visicol, ✦Enemol*): 1 adult or pediatric enema PR or 20-30 mL of oral soln PO prn (max 45 mL/24 h). Visicol: Evening before colonoscopy: 3 tabs with 8 oz clear liquid q15 min until 20 tabs are consumed. Day of colonoscopy: starting 3-5 h before procedure, 3 tabs with 8 oz clear liquid q15 min until 20 tabs are consumed. [OTC Trade only: pediatric & adult enema, oral solution. Rx Trade only: Visicol tab (trade $$$): 1.5 g.] ▶Not absorbed ♀C ▶? $

sorbitol: 30-150 mL (of 70% solution) PO or 120 mL (of 25-30% solution) PR as a single dose. Cathartic: 1-2 mL/kg PO. [Generic soln 70%] ▶Not absorbed ♀+ ▶? $

Laxatives - Stimulant

bisacodyl (*Correctol, Dulcolax, Feen-a-Mint*): 10-15 mg PO prn, 10 mg PR prn, 5-10 mg PR prn if 2-11 yo. [OTC Generic/Trade: tab 5 mg, supp 10 mg] ▶L ♀+ ▶? $

cascara: 325 mg PO qhs prn or 5 mL of aromatic fluid extract PO qhs prn. [OTC Generic: tab 325 mg, liquid aromatic fluid extract.] ▶L ♀C ▶+ $

castor oil (**Purge, Fleet Flavored Castor Oil**): 15-60 mL of castor oil or 30-60 mL emulsified castor oil PO qhs, 5-15 mL/dose of castor oil PO or 7.5-30 mL emulsified castor oil PO for child. [OTC Generic/Trade: liquid 30,60,120,480 mL, emulsified suspension 45,60,90,120 mL.] ▶Not absorbed ♀– ▶? $$$$

lubiprostone (**Amitiza**): 24 mcg PO bid with meals. [Trade only: 24 mcg caps.] ▶Gut ♀C ▶? $$$$

senna (**Senokot, SenokotXTRA, Ex-Lax, Fletcher's Castoria, ♣Glysennid**): 2 tabs or 1 tsp granules or 10-15 mL syrup PO. Max 8 tabs, 4 tsp granules, 30 mL syrup per day. Take granules with full glass of water. [OTC Generic/Trade (All dosing is based on sennosides content; 1 mg sennosides = 21.7 mg standardized senna concentrate): granules 15 mg/tsp, syrup 8.8 mg/5 mL, liquid 3 mg/mL (Fletcher's Castoria): tab 8.6, 15, 17, 25 mg , chewable tab 15 mg.] ▶L ♀C ▶+ $

Laxatives - Stool Softener

docusate calcium (**Surfak, Kaopectate Stool Softener**): 240 mg PO daily. [OTC Generic/Trade: cap 240 mg.] ▶L ♀+ ▶? $

docusate sodium (**Colace**): 50-500 mg/day PO divided in 1-4 doses. Peds: 10-40 mg/d if <3 yo, 20-60 mg/d if 3-6 yo, 40-150 mg/d if 6-12 yo. [OTC Generic/Trade: cap 50,100, 250 mg, tab 50,100 mg, liquid 10 & 50 mg/5 mL, syrup 16.75 & 20 mg/5 mL.] ▶L ♀+ ▶? $

Laxatives - Other or Combinations

mineral oil (**Kondremul, Fleet Mineral Oil Enema, ♣Lansoyl**): 15-45 mL PO. Peds: 5-15 mL/dose PO. Mineral oil enema: 60-150 mL PR. Peds 30-60 mL PR. [OTC Generic/Trade: plain mineral oil, mineral oil emulsion (Kondremul).] ▶Not absorbed ♀C ▶? $

Peri-Colace (docusate + sennosides): 2-4 tabs PO once daily or in divided doses prn. [OTC Generic/Trade: tab 50/8.6 mg docusate/sennosides.] ▶L ♀C ▶? $

Senokot-S (senna + docusate): 2 tabs PO daily. [OTC Generic/Trade: tab 8.6 mg senna concentrate/50 mg docusate.] ▶L ♀C ▶+ $

Other GI Agents

alosetron (**Lotronex**): Diarrhea-predominant IBS in women who have failed conventional therapy: 1 mg PO daily for 4 weeks; may increase to 1 mg PO bid. Discontinue if symptoms not controlled in 4 weeks on 1 mg PO bid. [Trade only: tab 0.5, 1 mg.] ▶L ♀B ▶? $$$$

alpha-galactosidase (**Beano**): 5 drops per ½ cup gassy food, 3 tabs PO (chew, swallow, crumble) or 15 drops per typical meal. [OTC Trade only: drops 150 GalU/5 drops, tab 150 GalU.] ▶Minimal absorption ♀? ▶? $

balsalazide (**Colazal**): 2.25 g PO tid x 8-12 weeks. [Trade only: cap 750 mg.] ▶Minimal absorption ♀B ▶? $$$$$

budesonide (**Entocort EC**): 9 mg PO daily x 8 weeks (remission induction) or 6 mg PO daily x 3 months (maintenance). [Trade only: cap 3 mg.] ▶L ♀C ▶? $$$$$

chlordiazepoxide-clidinium: 1 cap PO tid-qid. [Generic only: cap clidinium 2.5 mg + chlordiazepoxide 5 mg.] ▶K ♀D ▶- $

glycopyrrolate (**Robinul, Robinul Forte**): 0.1 mg/kg PO bid-tid, max 8 mg/day. [Trade only: tab 1, 2 mg.] ▶K ♀B ▶? $$$$

infliximab (**Remicade**): Moderately to severely active Crohn's disease, fistulizing disease, active ulcerative colitis (unresponsive to conventional therapies): 5 mg/ kg IV infusion at 0, 2, and 6 weeks, then q8 weeks. Serious, life-threatening infections, including sepsis & disseminated TB may occur, as can hypersensitivity reactions and hepatotoxicity. Monitor for CHF. ▶Serum ♀B ▶? $$$$$

lactase (*Lactaid*): Swallow or chew 3 caplets (Original strength), 2 caplets (Extra strength), 1 caplet (Ultra) with first bite of dairy foods. Adjust dose based on response. [OTC Generic/Trade: caplets, chew tab.] ▶Not absorbed ♀+ ▶+ $

Librax (chlordiazepoxide + methscopolamine): 1 cap PO tid-qid. [Trade only: cap methscopolamine 2.5 mg + chlordiazepoxide 5 mg.] ▶K ♀D ▶- $$$$$

mesalamine (**5-aminosalicylic acid, 5-ASA, Asacol, Pentasa, Canasa, Rowasa, ♣Mesasal, Salofalk**): Asacol: 800-1600 mg PO tid. Pentasa: 1000 mg PO qid. Canasa: 500 mg PR bid-tid or 1000 mg PR qhs. [Trade: delayed-release tab 400 mg (Asacol), controlled-release cap 250, 500 mg (Pentasa), rectal suppr 1000 mg (Canasa). Generic/Trade: rectal susp 4 g/60 mL (Rowasa).] ▶Gut ♀B ▶? $$$$$

neomycin (*Mycifradin, Neo-Fradin*): Hepatic encephalopathy: 4-12 g/day PO divided q6-8h. Peds: 50-100 mg/kg/day PO divided q6-8h. [Generic only: tab 500 mg. Trade (Neo-Fradin): solution 125 mg/5 mL.] ▶Minimally absorbed ♀D ▶? $$$

octreotide (*Sandostatin, Sandostatin LAR*): Variceal bleeding: Bolus 50-100 mcg IV followed by infusion 25-50 mcg/hr. AIDS diarrhea: 100-500 mcg SC tid. [Generic/Trade: injection vials 0.05, 0.1, 0.2, 0.5, 1 mg. Trade only: long-acting injectable susp (Sandostatin LAR) 10,20,30 mg.] ▶LK ♀B ▶? $$$$$

olsalazine (*Dipentum*): Ulcerative colitis: 500 mg PO bid. [Trade only: cap 250 mg.] ▶L ♀C ▶- $$$$

orlistat (*Xenical*): Weight loss: 120 mg PO tid with meals. [Trade only: Rx cap 120 mg, OTC cap 60 mg.] ▶Gut ♀B ▶? $$$$$

pancreatin (*Creon, Donnazyme, Ku-Zyme, ♣Entozyme*): 8,000-24,000 units lipase (1-2 tab/cap) PO with meals and snacks. [Tab, cap with varying amounts of pancreatin, lipase, amylase and protease.] ▶Gut ♀C ▶? $$$

pancrelipase (*Viokase, Pancrease, Pancrecarb, Cotazym, Ku-Zyme HP*): 4,000-33,000 units lipase (1-3 tab/cap) PO with meals and snacks. [Tab, cap, powder with varying amounts of lipase, amylase and protease.] ▶Gut ♀C ▶? $$$

pinaverine (*♣Dicetel*): Canada only. 50-100 mg PO tid. [Trade: tabs 50, 100 mg.] ▶? ♀C ▶- $$$

secretin (*SecreMax*): Test dose 0.2 mcg IV. If tolerated, 0.2-0.4 mcg/kg IV over 1 minute. ▶Serum ♀C ▶? $$$$$

sulfasalazine (*Azulfidine, Azulfidine EN-tabs, ♣Salazopyrin, Salazopyrin EN, S.A.S.*): 500-1000 mg PO qid. Peds: 30-60 mg/kg/day divided q4-6h. [Generic/Trade: tab 500 mg.] ▶K ♀- ▶? $$

tegaserod (*Zelnorm*): Constipation-predominant IBS in women: 6 mg PO bid before meals for 4-6 weeks. Chronic idiopathic constipation in those <65 yo: 6 mg PO bid before meals. [Trade only: tab 2, 6 mg.] ▶stomach/L ♀B ▶? $$$$

ursodiol (*Actigall, Ursofalk, URSO, URSO Forte*): Gallstone dissolution (Actigall): 8-10 mg/kg/day PO divided bid-tid. Prevention of gallstones associated with rapid weight loss (Actigall): 300 mg PO bid. Primary biliary cirrhosis (URSO): 13-15 mg/kg/day PO divided in 2-4 doses. [Generic/Trade: cap 300 mg. Trade only: tab 250 (URSO), 500 mg (URSO Forte).] ▶Bile ♀B ▶? $$$$

HEMATOLOGY (See cardiovascular section for antiplatelet drugs & thrombolytics.)

Anticoagulants - Heparin, LMW Heparins, & Fondaparinux

dalteparin (*Fragmin*): DVT prophylaxis, abdominal surgery: 2,500 units SC 1-2 h preop & daily postop x 5-10d. DVT prophylaxis, abdominal surgery in patients with malignancy: 5,000 units SC evening before surgery and daily postop x 5-10 days. DVT prophylaxis, hip replacement: Give SC for up to 14 days. Pre-op start: 2,500 units given 2 h preop and 4-8h postop, then 5,000 units daily starting ≥6h after second dose. Postop start: 2,500 units 4-8h postop, then 5,000 units daily starting ≥6h after first dose. Unstable angina or non-Q-wave MI: 120 units/kg up to 10,000 units SC q12h with aspirin (75-165 mg/day PO) until clinically stable. DVT prophylaxis, acute medical illness with restricted mobility: 5,000 units SC daily x 12-14 days. [Trade only: Single-dose syringes 2,500 & 5,000 anti-Xa units/0.2 mL, 7500 anti-Xa/0.3 mL, 10,000 anti-Xa/1 mL; multi-dose vial 10,000 units/mL, 9.5 mL and 25,000 units/mL, 3.8 mL.] ▶KL ♀B ▶+ $$$$

enoxaparin (*Lovenox*): DVT prophylaxis, hip/knee replacement: 30 mg SC q12h starting 12-24 h postop (CrCl <30 mL/min: 30 mg SC daily). Alternative for hip replacement: 40 mg SC daily starting 12h preop. Abdominal surgery: 40 mg SC daily starting 2 h preop (CrCl <30 mL/min: 30 mg SC daily). Acute medical illness with restricted mobility: 40 mg SC daily (CrCl <30 mL/min: 30 mg SC daily). Outpatient DVT treatment (no PE): 1 mg/kg SC q12h until PO anticoagulation established. Inpatient DVT treatment of (with/without PE): 1 mg/kg SC q12h or 1.5 mg/kg SC q24h (CrCl <30 mL/min: 1 mg/kg SC daily). Continue until therapeutic oral anticoagulation established. Unstable angina or non-Q-wave MI: 1 mg/kg SC q12h with aspirin (100-325 mg PO daily) for ≥2 days and until clinically stable (CrCl <30 mL/min: 1 mg/kg SC daily). [Trade only: Multi-dose vial 300 mg; Syringes 30,40 mg; graduated syringes 60,80,100,120, 150 mg. Concentration is 100 mg/mL except for 120,150 mg which are 150 mg/mL.] ▶KL ♀B ▶+ $$$$$

fondaparinux (*Arixtra*): DVT prophylaxis, hip/knee replacement or hip fracture surgery, abdominal surgery: 2.5 mg SC daily starting 6-8 h postop. Usual duration is 5-9 days; extend prophylaxis up to 24 additional days (max 32 days) in hip fracture surgery. DVT / PE treatment based on weight: 5 mg (if <50 kg), 7.5 mg (if 50-100 kg), 10 mg (if >100 kg) SC daily for ≥5 days & therapeutic oral anticoagulation. [Trade only: Pre-filled syringes 2.5 mg/0.5 mL, 5 mg/0.4 mL, 7.5 mg/0.6 mL, 10 mg/0.8 mL.] ▶K ♀B ▶? $$$$$

heparin (*Hepalean*): Venous thrombosis/pulmonary embolus treatment: Load 80 units/kg IV, then mix 25,000 units in 250 mL D5W (100 units/mL) and infuse at 18 units/kg/h. Adjust based on coagulation testing (PTT). DVT prophylaxis: 5,000 units SC q8-12h. Peds: Load 50 units/kg IV, then infuse 25 units/kg/h. [Generic: 1000, 2500, 5000, 7500, 10,000, 20,000 units/mL in various vial and syringe sizes.] ▶Reticuloendothelial system ♀C but + ▶+ $

WEIGHT-BASED HEPARIN DOSING FOR DVT/PE*

Initial dose: 80 units/kg IV bolus, then 18 units/kg/h. Check PTT in 6 h.
PTT <35 secs (<1.2 x control): 80 units/kg IV bolus, then ↑ infusion rate by 4 units/kg/h.
PTT 35-45 secs (1.2-1.5 x control): 40 units/kg IV bolus, then ↑ infusion by 2 units/kg/h.
PTT 46-70 seconds (1.5-2.3 x control): No change.
PTT 71-90 seconds (2.3-3 x control): ↓ infusion rate by 2 units/kg/h.
PTT >90 seconds (>3 x control): Hold infusion for 1 h, then ↓ infusion rate by 3 units/kg/h.

* PTT = Activated partial thromboplastin time. Reagent-specific target PTT may differ; use institutional nomogram when appropriate. Adjusted dosing may be appropriate in obesity. Consider establishing a max bolus dose / max initial infusion rate or use an adjusted body weight in obesity. Monitor PTT q6h during first 24h of therapy and 6h after each heparin dosage adjustment. The frequency of PTT monitoring can be reduced to q morning when PTT is stable within therapeutic range. Check platelets between days 3-5. Can begin warfarin on 1st day of heparin; continue heparin for ≥4 to 5 days of combined therapy. Adapted from *Ann Intern Med* 1993;119:874; *Chest* 2004:126:192S, *Circulation* 2001; 103:2994.

tinzaparin (**Innohep**): DVT with/without pulmonary embolism: 175 units/kg SC daily for ≥6 days and until adequate anticoagulation with warfarin. [Trade only: 20,000 anti-Xa units/mL, 2 mL multi-dose vial.] ▶K ♀B ▶+ $$$$$

Anticoagulants - Other

argatroban: Heparin-induced thrombocytopenia: Start 2 mcg/kg/min IV infusion. Get PTT at baseline and 2 h after starting infusion. Adjust dose (not >10 mcg/kg/min) until PTT is 1.5-3 times baseline (not >100 seconds). ▶L ♀B ▶- $$$$$

bivalirudin (**Angiomax**): Anticoagulation during PCI (including patients with or at risk of heparin-induced thrombocytopenia and heparin-induced thrombocytopenia and thrombosis syndrome: 0.75 mg/kg IV bolus prior to intervention, then 1.75 mg/kg/hr for duration of procedure (with provisional Gp IIb/IIIa inhibition). Use with aspirin 300-325 mg PO daily. Additional bolus of 0.3 mg/kg if activated clotting time <225 sec. ▶proteolysis/K ♀B ▶? $$$$$

lepirudin (**Refludan**): Anticoagulation in heparin-induced thrombocytopenia and associated thromboembolic disease: Bolus 0.4 mg/kg up to 44 mg IV over 15-20 seconds, then infuse 0.15 mg/kg/h up to 16.5 mg/h. Adjust dose to maintain APTT ratio of 1.5-2.5. ▶K ♀B ▶? $$$$$

warfarin (**Coumadin, Jantoven**): Start 2-10 mg PO daily x 3-4 days, then adjust dose to PT/INR. [Generic/Trade: Tabs 1, 2, 2.5, 3, 4, 5, 6, 7.5, 10 mg] ▶L ♀X ▶+ $

THERAPEUTIC GOALS FOR ANTICOAGULATION

INR Range*	Indication
2.0-3.0	Atrial fibrillation, deep venous thrombosis, pulmonary embolism, bioprosthetic heart valve, mechanical prosthetic heart valve (aortic position, bileaflet or tilting disk with normal sinus rhythm and normal left atrium)
2.5-3.5	Mechanical prosthetic heart valve: (1) mitral position, (2) aortic position with atrial fibrillation, (3) caged ball or caged disk

*Aim for an INR in the middle of the INR range (e.g. 2.5 for range of 2-3 and 3.0 for range of 2.5-3.5). Adapted from: Chest suppl 2004; 126: 416S, 450S, 474S; see this manuscript for additional information and other indications.

Other Hematological Agents (See endocrine section for vitamins and minerals.)

aminocaproic acid (**Amicar**): Hemostasis: 4-5 g PO/IV over 1h, then 1 g/h prn. [Generic/Trade: Syrup or oral soln 250 mg/mL, tabs 500 mg.] ▶K ♀D ▶? $$

anagrelide (**Agrylin**): Thrombocythemia due to myeloproliferative disorders: Start 0.5 mg PO qid or 1 mg PO bid, then after 1 week adjust to lowest effective dose. Max 10 mg/d. [Generic/Trade: Caps, 0.5, 1 mg.] ▶LK ♀C ▶? $$$$$

aprotinin (**Trasylol**): To reduce blood loss during CABG: 1 mL IV test dose ≥10 min before loading dose. Regimen A: 200 mL loading dose, then 200 mL pump prime dose, then 50 mL/h. Regimen B: 100 mL loading dose, then 100 mL pump prime dose, then 25 mL/h. May cause anaphylaxis. ▶lysosomal enzymes,K♀B▶? $$$$$

darbepoetin (**Aranesp, NESP**): Anemia of chronic renal failure: 0.45 mcg/kg IV/SC once weekly, or q2 weeks in some patients. Cancer chemo anemia: 2.25 mcg/kg SC q week, or 500 mcg SC every 3 weeks. Adjust dose based on Hb. [Trade only: Single-dose vials 25, 40, 60, 100, 200, 300, 500 mcg/1 mL. Additionally 150 mcg/0.75 mL. Prefilled syringes: 60, 100, 150, 200, 300, 500 mcg.] ▶cellular sialidases, L ♀C ▶? $$$$$

desmopressin (**DDAVP, Stimate, ✦Octostim**): Hemophilia A, von Willebrand's disease: 0.3 mcg/kg IV over 15-30 min, or 150-300 mcg intranasally. [Trade only:

Stimate nasal spray 150 mcg/0.1 mL (1 spray), 2.5 mL bottle (25 sprays). Generic/Trade (DDAVP nasal spray): 10 mcg/0.1 mL (1 spray), 5 mL bottle (50 sprays). Note difference in concentration of nasal solutions.] ▶LK ♀B ▶? $$$$$

erythropoietin (*Epogen, Procrit, epoetin, ✚Eprex*): Anemia: 1 dose IV/SC 3 times /week. Initial dose if renal failure = 50-100 units/kg, AZT = 100 units/kg, or chemo = 150 units/kg. Alternate for chemo-associated anemia: 40,000 units SC once/ wk. [Trade: Single-dose 1 mL vials 2,000, 3,000, 4,000, 10,000, 40,000 units/mL. Multi-dose vials 10,000 units/mL 2 mL & 20,000 units/mL 1 mL.] ▶L ♀C ▶? $$$$$

factor VIIa (*NovoSeven, ✚Niastase*): Specialized dosing. [Trade only: 1200, 2400, 4800 mcg/vial.] ▶L ♀C ▶? $$$$$

factor VIII (*Advate, Hemofil M, Humate P, Monoclate P, Monarc-M, ReFacto, ✚Kogenate, Recombinate*): Specialized dosing. [Specific formulation usually chosen by specialist in Hemophilia Treatment Center. Advate is only current recombinant formulation of factor VIII.] ▶L ♀C ▶? $$$$$

factor IX (*Benefix, Mononine, ✚Immunine VH*): Specialized dosing. [Specific formulation usually chosen by specialist in Hemophilia Treatment Center.] ▶L ♀C ▶? $$$$$

filgrastim (*G-CSF, Neupogen*): Neutropenia: 5 mcg/kg SC/IV daily. [Trade only: Single-dose vials 300 mcg/1 mL, 480 mcg/1.6 mL. Single-dose syringes 300 mcg/0.5 mL, 480 mcg/0.8 mL.] ▶L ♀C ▶? $$$$$

oprelvekin (*Neumega*): Chemotherapy-induced thrombocytopenia in adults: 50 mcg /kg SC daily. [Trade only: 5 mg single-dose vials with diluent.] ▶K ♀C ▶? $$$$$

pegfilgrastim (*Neulasta*): 6 mg SC once each chemo cycle. [Trade only: Single-dose syringes 6 mg/0.6 mL.] ▶Plasma ♀C ▶? $$$$$

protamine: Reversal of heparin: 1 mg antagonizes ~100 units heparin. Reversal of low molecular weight heparin: 1 mg protamine per 100 anti-Xa units of dalteparin or tinzaparin. 1 mg protamine per 1 mg enoxaparin. Give IV (max 50 mg) over 10 minutes. May cause allergy/anaphylaxis. ▶Plasma ♀C ▶? \$

sargramostim (*GM-CSF, Leukine*): Specialized dosing for marrow transplant. ▶L ♀C ▶? $$$$$

HERBAL & ALTERNATIVE THERAPIES

NOTE: In the US, herbal and alternative therapy products are regulated as dietary supplements, not drugs. Premarketing evaluation and FDA approval are not required unless specific therapeutic claims are made. Since these products are not required to demonstrate efficacy, it is unclear whether many of them have health benefits. In addition, there may be considerable variability in content from lot to lot or between products. See www.tarascon.com/herbals for the evidence-based efficacy ratings used by Tarascon editorial staff.

aloe vera (*acemannan, burn plant*): Topical: Efficacy unclear for seborrheic dermatitis, psoriasis, genital herpes, skin burns. Do not apply to surgical incisions; impaired healing reported. Oral: Mild to moderate active ulcerative colitis (possibly effective): 100 mL PO bid. Efficacy unclear for type 2 diabetes. OTC laxatives containing aloe removed from US market due to possible increased risk of colon cancer. ▶LK ♀oral- topical +? ▶oral- topical +? \$

androstenedione (*andro*): Marketed as anabolic steroid to enhance athletic performance. May cause androgenic (primarily in women) and estrogenic (primarily in men) side effects. FDA warned manufacturers to stop marketing as dietary supplement in 2004. ▶L, peripheral conversion to estrogens & androgens ♀- ▶- \$

aristolochic acid (*Aristolochia, Asarum, Bragantia*): Nephrotoxic & carcinogenic; do not use. Was promoted for weight loss. ▶? ♀- ▶- \$

arnica (*Arnica montana, leopard's bane, wolf's bane*): Do not take by mouth.

Topical promoted for treatment of skin wounds, bruises, aches, and sprains; but insufficient data to assess efficacy. Do not use on open wounds. ▶? ♀- ▶- $

artichoke leaf extract (*Cynara-SL, Cynara scolymus*): May reduce total cholesterol, but clinical significance is unclear. Cynara-SL is promoted as digestive aid (possibly effective for dyspepsia) at a dose of 1-2 caps PO daily (320 mg dried artichoke leaf extract/cap). ▶? ♀? ▶? $

astragalus (*Astragalus membranaceus, huang qi, vetch*): Used in combination with other herbs in traditional Chinese medicine, but efficacy unclear for CHD, CHF, chronic kidney disease, viral infections, and URIs. Possibly effective for improving survival and performance status with platinum-based chemotherapy for non-small cell lung cancer. ▶? ♀? ▶? $

bilberry (*Vaccinium myrtillus, huckleberry, Tegens, VMA extract*): Cataracts (efficacy unclear): 160 mg PO bid of 25% anthocyanosides extract. Insufficient data to evaluate efficacy for macular degeneration. Does not appear effective for improving night vision. ▶Bile, K ♀- ▶- $

bitter melon (*Momordica charantia, karela*): Possibly effective for type 2 diabetes. Dose unclear; juice may be more potent than dried fruit powder. Hypoglycemic coma reported in 2 children ingesting tea. Seeds can cause hemolytic anemia in G6PD deficiency. ▶? ♀- ▶- $$

bitter orange (*Citrus aurantium, Seville orange, Acutrim Natural AM, Dexatrim Natural Ephedrine Free*): Sympathomimetic similar to ephedra; safety and efficacy not established. Case reports of stroke and MI in patients taking bitter orange with caffeine. Do not use with MAOIs. ▶K ♀- ▶- $

black cohosh (*Cimicifuga racemosa, Remifemin, Menofem*): Menopausal symptoms (possibly effective): 20 mg PO bid of Remifemin. Efficacy unclear (conflicting data) for vasomotor symptoms induced by breast cancer treatment (including tamoxifen). North American Menopause Society considers black cohosh plus lifestyle changes an option for relief of mild symptoms. ▶? ♀- ▶- $

butterbur (*Petesites hybridus, Petadolex, Petaforce, Tesalin, ZE 339*): Migraine prophylaxis (possibly effective): Petadolex 50-75 mg PO bid. Allergic rhinitis prophylaxis (possibly effective): Petadolex 50 mg PO bid or Tesalin 1 tab PO qid or 2 tabs tid. Efficacy unclear for asthma or allergic skin disease. [Not by prescription. Standardized pyrrolizidine-free extracts: Petadolex (7.5 mg of petasin & isopetasin/50 mg tab). Tesalin (ZE 339; 8 mg petasin/tab).] ▶? ♀- ▶- $

chamomile (*Matricaria recutita - German chamomile, Anthemis nobilis - Roman chamomile*): Promoted as a sedative or anxiolytic, to relieve GI distress, for skin infections or inflammation, many other indications. Efficacy unclear for any indication. ▶? ♀- ▶? $

chaparral (*Larrea divaricata, creosote bush*): Hepatotoxic; do not use. Promoted as cancer cure. ▶? ♀- ▶- $

chasteberry (*Vitex agnus castus fruit extract, Femaprin*): Premenstrual syndrome (possibly effective): 20 mg PO daily of extract ZE 440. ▶? ♀- ▶- $

chondroitin: Efficacy for osteoarthritis is unclear (conflicting data). Glucosamine/Chondroitin Arthritis Intervention Trial (GAIT) did not find overall improvement in pain of knee OA with chondroitin 400 mg PO tid +/- glucosamine. Chondroitin + glucosamine improved pain in subgroup of patients with moderate to severe knee OA. ▶K ♀? ▶? $

coenzyme Q10 (*CoQ-10, ubiquinone*): Heart failure (efficacy unclear): 100 mg/day PO divided bid-tid. Parkinson's disease ($$$$): 1200 mg/day PO divided qid at meals and hs slowed progression of early disease in phase II study. Efficacy unclear for improving athletic performance. Appears ineffective for diabetes. Statins may reduce CoQ10 blood levels, but no evidence that CoQ10 supplements treat or prevent statin myopathy. ▶Bile ♀- ▶- $

comfrey (**Symphytum officinale**): May cause hepatic cancer; do not use, even topically. ▶? ♀- ▶- $

cranberry (**Cranactin, Vaccinium macrocarpon**): Prevention of UTI (possibly effective): 300 mL/day PO cranberry juice cocktail. Usual dose of cranberry juice extract caps/tabs is 300-400 mg PO bid. Insufficient data to assess efficacy for UTI treatment. Reports of increased INR and bleeding with warfarin. ▶? ♀- ▶? $

creatine: Promoted to enhance athletic performance. No benefit for endurance exercise; modest benefit for intense anaerobic tasks lasting <30 seconds. Usual loading dose of 20 g/day PO x 5 days, then 2-5 g/day taken bid. ▶LK ♀- ▶- $

dehydroepiandrosterone (**DHEA, Aslera, Fidelin, Prasterone**): No convincing evidence that DHEA slows aging or improves cognition in elderly. To improve wellbeing in women with adrenal insufficiency (effective): 50 mg PO daily. ▶Peripheral conversion to estrogens and androgens ♀- ▶- $

devil's claw (**Harpagophytum procumbens, Phyto Joint, Doloteffin, Harpadol**): Osteoarthritis, acute exacerbation of chronic low back pain (possibly effective): 2400 mg extract/day (50-100 mg harpagoside/day) PO divided bid-tid. Extracts standardized to harpagoside (iridoid glycoside) content. ▶? ♀- ▶- $

dong quai (**Angelica sinensis**): Appears ineffective for postmenopausal symptoms; North American Menopause Society recommends against use. May increase bleeding risk with warfarin; avoid concurrent use. ▶? ♀- ▶- $

echinacea (**E. purpurea, E. angustifolia, E. pallida, cone flower, EchinaGuard, Echinacin Madaus**): Efficacy unclear for prevention or treatment of upper respiratory infections. ▶? ♀- ▶- $

elderberry (**Sambucus nigra, Rubini, Sambucol, Sinupret**): Efficacy unclear for influenza, sinusitis, and bronchitis. ▶? ♀- ▶- $

ephedra (**Ephedra sinica, ma huang, Metabolife 356, Biolean, Ripped Fuel, Xenadrine**): Little evidence of efficacy, other than modest short-term weight loss. ▶K ♀- ▶- $

evening primrose oil (**Oenothera biennis**): Appears ineffective for premenstrual syndrome, postmenopausal symptoms, atopic dermatitis. ▶? ♀? ▶? $

fenugreek (**Trigonella foenum-graecum**): Efficacy unclear for diabetes or hyperlipidemia. ▶? ♀- ▶? $$

feverfew (**Chrysanthemum parthenium, Migra-Lief, MigraSpray, Tanacetum parthenium L.**): Prevention of migraine (possibly effective): 50-100 mg extract PO daily; 2-3 fresh leaves PO with or after meals daily; 50-125 mg freeze-dried leaf PO daily. May take 1-2 months to begin working. Inadequate data to evaluate efficacy for acute migraine. ▶? ♀- ▶- $

garcinia (**Garcinia cambogia, Citri Lean**): Appears ineffective for weight loss. ▶? ♀- ▶- $

garlic supplements (**Allium sativum, Kwai, Kyolic**): Modest reduction in lipids in short-term studies, but long-term benefit in hyperlipidemia unclear. Small reductions in BP, but efficacy in HTN unclear. Does not appear effective for diabetes. Cytochrome P450 3A4 inducer. Significantly decreases saquinavir levels. May increase bleeding risk with warfarin with/without increase in INR. ▶LK ♀- ▶- $

ginger (**Zingiber officinale**): Prevention of motion sickness (efficacy unclear): 500-1000 mg powdered rhizome PO single dose 1 h before exposure. American College of Obstetrics and Gynecology considers ginger 250 mg PO qid a nonpharmacologic option for N/V of pregnancy. Efficacy unclear for postop N/V (conflicting study results). ▶? ♀? ▶? $

ginkgo biloba (*EGb 761, Ginkgold, Ginkoba, Quanterra Mental Sharpness*): Dementia (modestly effective): 40 mg PO tid of standardized extract containing 24% ginkgo flavone glycosides and 6% terpene lactones. Benefit may be delayed for up to 4 weeks. Does not appear to improve memory in elderly with normal cognitive function. Does not appear effective for prevention of acute altitude sickness. Limited benefit in intermittent claudication. ▶K ♀- ▶- $

ginseng - American (*Panax quinquefolius L.*): Reduction of postprandial glucose in type 2 diabetes (possibly effective): 3 g PO taken with or up to 2h before meal. ▶K ♀- ▶- $

ginseng - Asian (*Panax ginseng, Ginsana, Ginsai, G115, Korean red ginseng*): Promoted to improve vitality and well-being: 200 mg PO daily. Ginsana: 2 caps PO daily or 1 cap PO bid. Ginsana Sport: 1 cap PO daily. Preliminary evidence of efficacy for erectile dysfunction. Efficacy unclear for improving physical or psychomotor performance, diabetes, herpes simplex infections, cognitive or immune function, postmenopausal hot flashes (American College of OB/GYN and North American Menopause Society recommend against use). ▶? ♀- ▶- $

ginseng - Siberian (*Eleutherococcus senticosus. Ci-wu-jia*): Does not appear effective for improving athletic endurance, or chronic fatigue syndrome. May interfere with FPIA and MEIA digoxin assays. ▶? ♀- ▶- $

glucosamine (*Aflexa, Cosamin DS, Dona, Flextend, Promotion*): Efficacy for osteoarthritis is unclear (conflicting data). Glucosamine/ Chondroitin Arthritis Intervention Trial (GAIT) did not find overall improvement in pain of knee OA with glucosamine HCl 500 mg +/- chondroitin 400 mg both PO tid. Glucosamine + chondroitin did improve pain in subgroup of patients with moderate to severe OA. Some earlier studies reported improved pain with a different salt [glucosamine sulfate (Dona) 1500 mg PO once daily.] ▶? ♀- ▶- $

goldenseal (*Hydrastis canadensis*): Often used in attempts to achieve false-negative urine test for illicit drug use (efficacy unclear). Often combined with echinacea in cold remedies; but insufficient data to assess efficacy for common cold or URIs. ▶? ♀- ▶- $

grape seed extract (*Vitus vinifera L., procyanidolic oligomers, PCO*): Small clinical trials suggest benefit in chronic venous insufficiency. No benefit in single study of seasonal allergic rhinitis. ▶? ♀? ▶? $

green tea (*Camellia sinensis*): Efficacy unclear for cancer prevention, weight loss, hypercholesterolemia. Do not use in patients receiving irinotecan. Large doses might decrease INR with warfarin due to vitamin K content. Contains caffeine. [Green tea extract available in caps standardized to polyphenol content.] ▶? ♀+ in moderate amount in food, - in supplements $

guarana (*Paullinia cupana*): Marketed as an ingredient in weight-loss dietary supplements. Seeds contain caffeine. Guarana in weight loss dietary supplements has the potential to provide high doses of caffeine. ▶? ♀+ in food, - in supplements ▶+ in food, - in supplements ?

guggulipid (*Commiphora mukul extract, guggul*): Efficacy unclear for treatment of hyperlipidemia (conflicting study results). ▶? ♀- ▶- $$

hawthorn (*Crataegus laevigata, monogyna, oxyacantha, standardized extract WS 1442 - Crataegus novo, HeartCare*): Mild heart failure (possibly effective): 80 mg PO bid to 160 mg PO tid of standardized extract (19% oligomeric procyanidins; WS 1442; HeartCare 80 mg tabs). ▶? ♀- ▶- $$

horse chestnut seed extract (*Aesculus hippocastanum, HCE50, Venastat*): Chronic venous insufficiency (effective): 1 cap Venastat (16% aescin standardized extract) PO bid with water before meals. ▶? ♀- ▶- $

kava (***Piper methysticum, One-a-day Bedtime & Rest, Sleep-Tite***): Promoted as anxiolytic (possibly effective) or sedative. Do not use due to hepatotoxicity. ▶K ♀- ▶- $

kombucha tea (***Manchurian or Kargasok tea***): Recommend against use; has no proven benefit for any indication; may cause severe acidosis. ▶? ♀- ▶- $

licorice (***Glycyrrhiza glabra, Glycyrrhiza uralensis***): Insufficient data to assess efficacy for postmenopausal vasomotor symptoms. Chronic high doses can cause pseudo-primary aldosteronism (with HTN, edema, hypokalemia). ▶Bile ♀- ▶- $

melatonin (***N-acetyl-5-methoxytryptamine***): To reduce jet lag after flights over >5 time zones (possibly effective): 0.5-5 mg PO qhs x 3-6 nights starting on day of arrival. ▶L ♀- ▶- $

methylsulfomethane (***MSM, dimethyl sulfone, crystalline DMSO2***): Insufficient data to assess efficacy of oral and topical MSM for arthritis pain. ▶? ♀- ▶- $

milk thistle (***Silybum marianum, Legalon, silymarin, Thisylin***): Hepatic cirrhosis (possibly effective): 100-200 mg PO tid of standardized extract with 70-80% silymarin. ▶LK ♀- ▶- $

nettle root (***stinging nettle, Urtica dioica radix***): Efficacy unclear for treatment of BPH. ▶? ♀- ▶- $

noni (***Morinda citrifolia***): Promoted for many disorders; but insufficient data to assess efficacy. Potassium content comparable to orange juice; hyperkalemia reported in chronic renal failure. Case reports of hepatotoxicity. ▶? ♀- ▶- $$$

policosanol (***One-A-Day Cholesterol Plus, CholeRx, Cholest Response, Cholestin***): Promoted for hyperlipidemia (efficacy unclear). A Cuban formulation (unavailable in US) reduced LDL cholesterol in studies by a single group of researchers, but studies by other groups found no benefit. US formulations have not been evaluated in clinical trials. ▶? ♀- ▶- $

probiotics (***Acidophilus, Bifidobacteria, Lactobacillus, Bacid, Culturelle, Florastor, IntestiFlora, Lactinex, LiveBac, Power-Dophilus, Primadophilus, Probiotica, Saccharomyces boulardii, VSL#3***): Prevention of antibiotic-associated diarrhea (effective): Forastor (Saccharomyces boulardii) 2 caps PO bid for adults; 1 cap PO bid for peds. Culturelle (Lactobacillus GG) 1 cap PO once daily or bid for peds. Give 2 h before/after antibiotic. Peds rotavirus gastroenteritis (effective): Lactobacillus GG ≥10 billion cells/day PO started early in illness. Prevention of recurrent C difficile diarrhea in adults (possibly effective): Florastor 2 caps PO bid x 4 weeks. VSL#3 (approved as medical food) for ulcerative colitis or pouchitis: 1-4 packets/day for adults depending on number of bowel movements; peds dose based on weight and number of bowel movements. [Not by prescription. Culturelle contains Lactobacillus GG 10 billion cells/cap. Florastor contains Saccharomyces boulardii 5 billion cells/250 mg cap. Probiotica contains Lactobacillus reuteri 100 million cells/chew tab. VSL#3 contains 450 billion cells/packet (Bifidobacterium breve, longum, infantis; Lactobacillus acidophilus, plantarum, casei, bulgaricus; Streptococcus thermophilus).] ▶? ♀+ ▶+ $

pycnogenol (***French maritime pine tree bark***): Promoted for many medical disorders; but insufficient data to assess efficacy. ▶L ♀? ▶? $

pygeum africanum (***African plum tree, Prostata, Prostatonin, Provol***): BPH (may have modest efficacy): 50-100 mg PO bid or 100 mg PO daily of standardized extract containing 14% triterpenes. Prostatonin (also contains Urtica dioica): 1 cap PO bid with meals; up to 6 weeks for full response. ▶? ♀- ▶- $

red clover isoflavone extract (***Trifolium pratense, trefoil, Promensil, Rimostil, Supplifem, Trinovin***): Postmenopausal vasomotor symptoms (conflicting evi-

..., does not appear effective overall, but may have modest benefit for severe .... Promensil 1 tab PO daily-bid with meals. [Not by prescription. Isoflavone content (genistein, daidzein, biochanin, formononetin) is 40 mg/tab in Promensil and Trinovin, 57 mg/tab in Rimostil.] ►Gut, L, K ♀- ▶- $$

red yeast rice (*Monascus purpureus*): Efficacy of currently available products for hyperlipidemia is unclear. A formulation of Cholestin was pulled from the US market because it contained a small amount of lovastatin formed in the fermentation process. The current Cholestin formulation contains policosanol (efficacy unclear; conflicting evidence). It is unclear whether current red yeast rice products contain statins. ►L ♀- ▶- $$

s-adenosylmethionine (*SAM-e, sammy*): Depression (possibly effective): 400-1600 mg/day PO. Osteoarthritis (possibly effective): 400-1200 mg/day PO. Onset of response in OA in 2-4 weeks. ►L ♀? ▶? $$$

Saint John's wort (*Alterra, Hypericum perforatum, Kira, Movana, One-a-day Tension & Mood, LI-160, St John's wort*): Mild depression (effective): 300 mg PO tid of standardized extract (0.3% hypericin). May be ineffective for moderate major depression. May decrease efficacy of many drugs (eg, oral contraceptives) by inducing liver metabolism. May cause serotonin syndrome with SSRIs, MAOIs. ►L ♀- ▶- $

saw palmetto (*Serenoa repens, One-a-day Prostate Health, Quanterra*): BPH (possibly effective for mild to moderate; appears ineffective for moderate to severe): 160 mg PO bid or 320 mg PO daily of standardized liposterolic extract. Take with food. Brewed teas may not be effective. ▶? ♀- ▶- $

shark cartilage (*BeneFin, Cancenex, Cartilade*): Appears ineffective for palliative care of advanced cancer. ▶? ♀- ▶- $$$$$

silver - colloidal (*mild & strong silver protein, silver ion*): Promoted as antimicrobial; unsafe and ineffective for any use. Silver accumulates in skin (leads to grey tint), conjunctiva, and internal organs with chronic use. [Not by prescription. May come as silver chloride, cyanide, iodide, oxide, or phosphate.] ▶? ♀- ▶- $

soy (*Genisoy, Healthy Woman, Novasoy, Phytosoya, Supplfem, Supro*): Cardiovascular risk reduction: ≥25 g/day soy protein (50 mg/day isoflavones) PO. Hypercholesterolemia: ~50 g/day soy protein PO reduces LDL cholesterol by ~3%; no apparent benefit for isoflavone supplements. Postmenopausal vasomotor symptoms (conflicting evidence; modest benefit if any): 20-60 g/day soy protein PO (40-80 mg/day isoflavones). Conflicting evidence for postmenopausal bone loss. ►Gut, L, K ♀+ for food, ▶ for supplements ▶- for food, ▶ for supplements $

stevia (*Stevia rebaudiana*): Leaves traditionally used as sweetener. Efficacy unclear for treatment of type 2 diabetes or hypertension. ►L ♀? ▶? $

tea tree oil (*melaleuca oil, Melaleuca alternifolia*): Not for oral use; CNS toxicity reported. Efficacy unclear for onychomycosis, tinea pedis, acne vulgaris, dandruff. ▶? ♀- ▶- $

valerian (*Valeriana officinalis, Alluna, One-a-day Bedtime & Rest, Sleep-Tite*): Insomnia (possibly effective): 400-900 mg of standardized extract PO 30 minutes before bedtime. Alluna: 2 tabs PO 1 h before bedtime. ▶? ♀- ▶- $

wild yam (*Dioscorea villosa*): Ineffective as topical "natural progestin". Was used historically to synthesize progestins, cortisone, and androgens; it is not converted to them or DHEA in the body. ►L ♀? ▶? $

willow bark extract (*Salix alba, Salicis cortex, Assalix, salicin*): Osteoarthritis, low back pain (possibly effective): 60 to 240 mg/day salicin PO divided bid-tid. [Not by prescription. Some products standardized to 15% salicin content.] ►K ♀- ▶- $

yohimbe (*Corynanthe yohimbe, Pausinystalia yohimbe, Potent V*): Nonprescription yohimbe promoted for impotence and as aphrodisiac, but these products rarely contain much yohimbine. FDA considers yohimbe bark in herbal remedies an unsafe herb. [Yohimbine is the primary alkaloid in the bark of the yohimbe tree. Yohimbine HCl is a prescription drug in the US; yohimbe bark is available without prescription. Yohimbe bark (non prescription) and prescription yohimbine HCl are not interchangeable.] ▶L ♀- ▶- $

IMMUNOLOGY

Immunizations (For vaccine info see CDC website www.cdc.gov.)

BCG vaccine (*Tice BCG*, ♥*Pacis, Oncotice, Immucyst*): 0.2-0.3 mL percutaneously. ♀C ▶? $$$$

Comvax (haemophilus b + hepatitis B vaccine): 0.5 mL IM. ♀C ▶? $$$

diphtheria tetanus & acellular pertussis vaccine (*DTaP, Tripedia, Infanrix, Daptacel, Boostrix, Adacel*, ♥*Tripacel*): 0.5 mL IM. ♀C ▶- $$

diphtheria-tetanus toxoid (*Td, DT*, ♥*D2T5*): 0.5 mL IM. [Injection DT (pediatric: 6 weeks-6 yo). Td (adult and children: ≥7 years).] ♀C ▶? $

haemophilus b vaccine (*ActHIB, HibTITER, PedvaxHIB*): 0.5 mL IM. ♀C ▶? $

hepatitis A vaccine (*Havrix, Vaqta*, ♥*Avaxim, Epaxal*): Adult formulation 1 mL IM, repeat in 6-12 months. Peds ≥1 yo: 0.5 mL IM, repeat 6-18 months later. [Single dose vial (specify pediatric or adult).] ♀C ▶+ $$$

hepatitis B vaccine (*Engerix-B, Recombivax HB*): Adults: 1 mL IM, repeat in 1 and 6 months. Separate pediatric formulations and dosing. ♀C ▶+ $$$

human papillomavirus recombinant vaccine (*Gardasil*): 0.5 mL IM at 0, 2 and 6 months. ♀B ▶? $$$$

influenza vaccine (*Fluarix, Fluzone, Fluvirin, FluMist*, ♥*Fluviral, Vaxigrip*): 0.5 mL IM or 1 dose (0.5 mL) intranasally (FluMist). Fluarix not for children. ♀C ▶+ $

Japanese encephalitis vaccine (*JE-Vax*): 1.0 mL SC x 3 doses on days 0, 7, and 30. ♀C ▶? $$$$

measles mumps & rubella vaccine (*M-M-R II*, ♥*Priorix*): 0.5 mL SC. ♀C ▶+ $$

meningococcal polysaccharide vaccine (*Menomune-A/C/Y/W-135, Menactra*, ♥*Menjugate*): 0.5 mL SC (Menomune) or IM (Menactra). ♀C ▶? $$$

CHILDHOOD IMMUNIZATION SCHEDULE*						Months				Years	
Age	Birth	1	2	4	6	12	15	18	24	4-6	11-12
Hepatitis B	HB-1		HB-2			HB-3					
DTP†			DTP	DTP	DTP		DTP			DTP	DTaP
H influenza b			Hib	Hib	Hib	Hib					
Pneumococci			PCV	PCV	PCV	PCV					
Polio§			IPV	IPV		IPV				IPV	
MMR						MMR				MMR	
Varicella						Varicella					
Meningococcal											MCV
Influenza						Influenza‡					
Hepatitis A¶										HA (some areas)	

*2006 schedule from the CDC, ACIP, AAP, & AAFP, see CDC website (www.cdc.gov). †Acellular form preferred for all DTP doses during primary immunization series. DTaP (adolescent preparation) recommended at 11-12 years. Also annually immunize against influenza, older children (and contacts) with risk factors such as asthma, cardiac disease, sickle cell diseases, HIV, DM. Live-attenuated influenza (nasal form) only for healthy children ≥5 yo. ¶HA series recommended for selected high-risk areas, consult local public health authorities.

)TaP + hepatitis B + polio): 0.5 mL at 2, 4, 6 months IM. ♀C ▷? $$$

...cine (**Plague vaccine**): Age 18-61 yo: 1 mL IM x 1 dose, then 0.2 mL IM 1-3 months after the 1st injection, then 0.2 mL IM 5-6 months later. ♀C ▷+ $

pneumococcal 23-valent vaccine (**Pneumovax, ✿ Pneumo 23**): 0.5 mL IM/SC. ▷+ $$

pneumococcal 7-valent conjugate vaccine (**Prevnar**): 0.5 mL IM x3 doses 6-8 wks apart starting at 2-6 months of age, then 4th dose at 12-15 months. ♀C ▷? $$$

poliovirus vaccine (**IPOL**): 0.5 mL IM or SC. ♀C ▷+ $$

ProQuad (measles mumps & rubella vaccine + varicella vaccine): 12 months-12 years: 0.5 mL (1 vial) SC. ♀C ▷? $$$$

rabies vaccine (**RabAvert, Imovax Rabies, BioRab, Rabies Vaccine Adsorbed**): 1 mL IM in deltoid region on days 0, 3, 7, 14, 28. ♀C ▷? $$$$$

rotavirus vaccine (**RotaTeq**): Give the first dose (2 mL PO) between 6-12 weeks of age, and then the 2nd & 3rd doses at 4-10 weeks intervals thereafter (last dose no later than 32 weeks). [Trade only: Oral suspension 2 mL.] ♀ ▷? $$$$$

tetanus toxoid: 0.5 mL IM/SC. ♀C ▷+ $

TriHibit (haemophilus b + DTaP): 4th dose only, 15-18 mos: 0.5 mL IM. ♀C ▷- $$

Twinrix (hepatitis A inactivated + hepatitis B recombinant vaccines): Adults: 1 mL IM in deltoid, repeat in 1 & 6 months. ♀C ▷? $$$$

typhoid vaccine (**Vivotif Berna, Typhim Vi, ✿ Typherix**): 0.5 mL IM x 1 dose (Typhim Vi); 1 cap qod x 4 doses (Vivotif Berna). May revaccinate q2-5 yrs if high risk. [Trade only: Caps] ♀C ▷? $$$

varicella vaccine (**Varivax, ✿ Varilrix**): Children 1 to 12 yo: 0.5 mL SC x 1 dose. Age ≥13: 0.5 mL SC, repeat 4-8 weeks later. ♀C ▷+ $$$$

yellow fever vaccine (**YF-Vax**): 0.5 mL SC. ♀C ▷+ $$$

zoster vaccine - live (**Zostavax**): Adults ≥60 yo: 0.65 mL SC x 1. ♀C ▷? $$$$

Immunoglobulins

antivenin - crotalidae immune Fab ovine polyvalent (**CroFab**): Rattlesnake envenomation: Give 4-6 vials IV infusion over 60 minutes, within 6 hours of bite if possible. Administer 4-6 additional vials if no initial control of envenomation syndrome, then 2 vials q6h for up to 18 hours (3 doses) after initial control has been established. ▷? ♀C ▷? $$$$$

botulism immune globulin (**BabyBIG**): Infant botulism <1 yo: 1 mL (50 mg)/kg IV. ▷L ♀? ▷? $$$$$

hepatitis B immune globulin (**H-BIG, BayHep B, HepaGam, NABI-HB**): 0.06 mL/kg IM within 24 h of needlestick, ocular, or mucosal exposure, repeat in 1 month. ▷L ♀C ▷? $$$

immune globulin - intramuscular (**Baygam**): Hepatitis A prophylaxis: 0.02-0.06 mL/kg IM depending on length of stay. Measles (within 6 days post-exposure): 0.2-0.25 mL/kg IM. ▷L ♀C ▷? $$$$

immune globulin - intravenous (**Carimune, Gamimune, Polygam, Panglobulin, Octagam, Flebogamma, Sandoglobulin, Gammagard, Gamunex, Iveegam, Venoglobulin**): IV dosage varies by indication and product. ▷L ♀C ▷? $$$$$

immune globulin - subcutaneous (**Vivaglobulin**): 100-200 mg/kg SC weekly. ▷L ♀C ▷? $$$$$

lymphocyte immune globulin human (**Atgam**): Specialized dosing. ▷L ♀C ▷? $$$$$

rabies immune globulin human (**Imogam, BayRab**): 20 units/kg, as much as possible infiltrated around bite, the rest IM. ▷L ♀C ▷? $$$$$

RSV immune globulin (**RespiGam**): IV infusion for RSV. ▷Plasma ♀C ▷? $$$$$

tetanus immune globulin (*BayTet*): Prophylaxis: 250 units IM. ▶L ♀C ▶? $$$$
varicella-zoster immune globulin (*VZIG*): Specialized dosing. ▶L ♀C ▶? $$$$$

TETANUS WOUND CARE (www.cdc.gov)	Uncertain or <3 prior tetanus immunizations	≥3 prior tetanus immunizations
Non tetanus prone wound (e.g., clean and minor)	Td (DT if <7 yo)	Td if >10 years since last dose
Tetanus prone wound (e.g., dirt, contamination, punctures, crush components)	Td (DT if <7 yo), tetanus immune globulin 250 units IM at site other than Td.	Td if >5 years since last dose

When immunizing adults or adolescents ≥10 yo consider DTaP (*Adacel* if 11-64 yo, *Boostrix* if 10-18 yo) if patient has never received a pertussis booster.

Immunosuppression (Specialized dosing for organ transplantation.)

basiliximab (*Simulect*): ▶Plasma ♀B ▶? $$$$$
cyclosporine (*Sandimmune, Neoral, Gengraf*): Specialized dosing for organ transplantation, rheumatoid arthritis, and psoriasis. [Generic/Trade: microemulsion Caps 25, 100 mg. Generic/Trade: Caps (Sandimmune) 25, 100 mg, solution (Sandimmune) 100 mg/mL, microemulsion solution (Neoral, Gengraf) 100 mg/mL.] ▶L ♀C ▶- $$$$$
daclizumab (*Zenapax*): ▶L ♀C ▶? $$$$$
mycophenolate mofetil (*Cellcept, Myfortic*): [Trade only (CellCept): caps 250 mg, tabs 500 mg, oral suspension 200 mg/mL. Trade (Myfortic): tablet, extended-release: 180, 360 mg.] ▶? ♀C ▶? $$$$$
sirolimus (*Rapamune*): [Trade: oral soln 1 mg/mL. Tab 1,2 mg.] ▶L ♀C ▶- $$$$$
tacrolimus (*Prograf, FK 506*): [Trade only: Caps 1,5 mg.] ▶L ♀C ▶- $$$$$

Other

tuberculin PPD (*Aplisol, Tubersol, Mantoux, PPD*): 5 TU (0.1 mL) intradermally, read 48-72h later. ▶L ♀C ▶+ $

NEUROLOGY

Alzheimer's Disease - Cholinesterase Inhibitors

donepezil (*Aricept*): Start 5 mg PO qhs. May increase to 10 mg PO qhs in 4-6 wks. [Trade only: Tabs 5,10 mg. Orally disintegrating tabs 5,10 mg.] ▶LK ♀C ▶? $$$$
galantamine (*Razadyne, Razadyne ER*): Extended release: Start 8 mg PO q am with food; increase to 16 mg q am after 4 wks. May increase to 24 q am after another 4 wks. Immediate release: Start 4 mg PO bid with food; increase to 8 mg bid after 4 wks. May increase to 12 mg bid after another 4 wks. [Trade only: Razadyne tabs 4, 8, 12 mg; oral solution 4 mg/mL. Razadyne ER extended release caps 8, 16, 24 mg. Prior to April 2005 was called Reminyl.] ▶LK ♀B ▶? $$$$
rivastigmine (*Exelon*): Alzheimer's disease: Start 1.5 mg PO bid with food. Increase to 3 mg bid after 2 wks. Max 12 mg/d. Parkinson's dementia: Start 1.5 mg PO bid with food. Increase by 3 mg/d at intervals >4 weeks to max 12 mg/d. [Trade only: Caps 1.5, 3, 4.5, 6 mg. Oral solution 2 mg/mL (120 mL).] ▶K ♀B ▶? $$$$

Alzheimer's Disease - NMDA Receptor Antagonists

memantine (*Namenda*, ✦*Ebixa*): Start 5 mg PO daily. Increase by 5 mg/d at weekly intervals to max 20 mg/d. Doses >5 mg/d should be divided bid. [Trade only: Tabs 5, 10 mg. Oral soln 2 mg/mL.] ▶KL ♀B ▶? $$$$

Anticonvulsants

carbamazepine (**Tegretol, Tegretol XR, Carbatrol, Epitol**): 200-400 mg PO bid-qid. Extended-release: 200 mg PO bid. Age 6-12 yo: 100 mg PO bid or 50 mg PO qid; increase by 100 mg/d at weekly intervals divided tid-qid (regular release), bid (extended-release), or qid (suspension). Age <6 yo: 10-20 mg/kg/d PO divided bid-qid. Aplastic anemia. [Generic/Trade: Tabs 200 mg, chew tabs 100 mg, susp 100 mg/5 ml. Generic Only: Tabs 100,300,400 mg, chew tabs 200 mg. Trade only: Extended-release tabs (Tegretol XR): 100, 200, 400 mg. Extended-release caps (Carbatrol): 100, 200, 300 mg.] ▶LK ♀D ▶+ $$

clobazam (✦**Frisium**): Canada only. Adults: Start 5-15 mg PO daily. Increase prn to max 80 mg/d. Children <2 yo: 0.5-1 mg/kg PO daily. Children 2-16 yo: Start 5 mg PO daily. May increase prn to max 40 mg/d. [Generic/Trade: Tabs 10 mg.] ▶L ♀X (first trimester) D (2nd/3rd trimesters) ▶- $

clonazepam (**Klonopin, Klonopin Wafer, ✦Clonapam, Rivotril**): Start 0.5 mg PO tid. Max 20 mg/d. [Generic/Trade: Tabs 0.5, 1, 2 mg. Orally disintegrating tabs (approved for panic disorder only) 0.125, 0.25, 0.5, 1, 2 mg.] ▶LK ♀D ▶- ⊙IV $$$

diazepam (**Valium, Diastat, Diastat AcuDial, ✦Vivol, E Pam, Diazemuls**): Active seizures: 5-10 mg IV q10-15 min to max 30 mg, or 0.2-0.5 mg/kg rectal gel PR. Muscle spasm: 2-10 mg PO tid-qid. [Generic/Trade: Tabs 2, 5, 10 mg. Soln 5 mg/5 mL. Trade only: Intensol concentrated soln 5 mg/mL. Rectal gel (Diastat) 2.5, 5, 10, 15, 20 mg. Rectal gel (Diastat AcuDial) 10, 20 mg.] ▶LK ♀D ▶- ⊙IV $$$

ethosuximide (**Zarontin**): Start 250 mg PO daily (or divided bid) if 3-6 yo, or 500 mg PO daily (or divided bid) if >6 yo. Max 1.5 g/d. [Generic/Trade: Caps 250 mg. Syrup 250 mg/5 mL.] ▶L ♀C ▶+ $$$$

felbamate (**Felbatol**): Start 400 mg PO tid. Max 3,600 mg/d. Peds: Start 15 mg/kg/day PO divided tid-qid. Max 45 mg/kg/d. Aplastic anemia, hepatotoxicity. [Trade only: Tabs 400, 600 mg. Susp 600 mg/5 mL.] ▶KL ♀C ▶- $$$$$

fosphenytoin (**Cerebyx**): Load: 15-20 mg "phenytoin equivalents" (PE) per kg IM/IV no faster than 150 mg/min. Maintenance: 4-6 PE/kg/d. ▶L ♀D ▶+ $$$$$

gabapentin (**Neurontin**): Start 300 mg PO qhs. Increase gradually to 300-600 mg PO tid. Max 3,600 mg/d. Postherpetic neuralgia: Start 300 mg PO on day 1; increase to 300 mg bid on day 2, and to 300 mg tid on day 3. Max 1,800 mg/d. Partial seizures, initial monotherapy: Titrate as for other indications. Usual effective dose is 900-1,800 mg/day. [Generic: Tabs 100,300,400 mg. Generic/Trade: Caps 100, 300, 400 mg. Tabs 600, 800 mg (scored). Soln 50 mg/mL.] ▶K ♀C ▶? $$$$

lamotrigine (**Lamictal, Lamictal CD**): Partial seizures or Lennox-Gastaut syndrome, adjunctive therapy with a single enzyme-inducing anticonvulsant. Age >12 yo: 50 mg PO daily x 2 wks, then 50 mg bid x 2 wks, then gradually increase to 150-250 mg PO bid. Age 2-12 yo: dosing is based on weight and concomitant meds (see package insert). Also approved for conversion to monotherapy (age ≥16 yo): see package insert. Drug interaction with valproate (see package insert for adjusted dosing guidelines). Potentially life-threatening rashes reported in 0.3% of adults and 0.8% of children; discontinue at first sign of rash. [Generic/Trade: Chewable dispersible tabs 5, 25 mg. Trade only: Tabs 25, 100, 150, 200 mg. Chewable dispersible tabs (Lamictal CD) 2 mg not available in pharmacies.] ▶LK ♀C ▶- $$$$

levetiracetam (**Keppra**): Partial seizures, juvenile myoclonic epilepsy (JME): Start 500 mg PO bid; increase by 1,000 mg/d q2 wks prn to max 3,000 mg/d (partial seizures) or to target dose of 3,000 mg/d (JME). [Trade only: Tabs 250, 500, 750, 1000 mg. Oral solution 100 mg/mL.] ▶K ♀C ▶? $$$$$

lorazepam (**Ativan**): Status epilepticus (unapproved): 0.05-0.1 mg/kg IV over 2-5 min. ▶LK ♀D ▶- ⊙IV $$$

oxcarbazepine (***Trileptal***): Start 300 mg PO bid. Titrate to 1200 mg/d (adjunctive) or 1,200-2,400 mg/d (monotherapy). Peds 2-16 yo: Start 8-10 mg/kg/d divided bid. [Trade: Tabs (scored) 150, 300, 600 mg. Susp 300 mg/5 mL.] ▶LK ♀C ▶- $$$$$

phenobarbital (***Luminal***): Load: 20 mg/kg IV at rate ≤60 mg/min. Maintenance: 100-300 mg/d PO given once daily or divided bid; peds 3-5 mg/kg/day PO divided bid-tid. Multiple drug interactions. [Generic/Trade: Tabs 15, 16, 30, 32, 60, 65, 100 mg. Elixir 20 mg/5 mL.] ▶L ♀D ▶- ©IV $

phenytoin (***Dilantin, Phenytek***): Status epilepticus: Load 10-15 mg/kg IV no faster than 50 mg/min, then 100 mg IV/PO q6-8h. Epilepsy: Oral load: 400 mg PO initially, then 300 mg in 2h and 4h. Maintenance: 5 mg/kg (or 300 mg PO) given daily (extended-release) or divided tid (standard release). [Generic/Trade: Extended-release caps 100 mg (Dilantin). Suspension 125 mg/5 mL. Trade only: Extended-release caps 30 mg (Dilantin), 200, 300 mg (Phenytek). Chew tabs 50 mg. Generic: Prompt-release caps 100 mg.] ▶L ♀D ▶+ $$

pregabalin (***Lyrica***): Painful diabetic peripheral neuropathy: Start 50 mg PO tid; may increase after 1 wk to max 100 mg PO tid. Postherpetic neuralgia: Start 150 mg/d PO divided bid-tid. May increase after 1 wk to 300 mg/d divided bid-tid; max 600 mg/d. Partial seizures (adjunctive): Start 150 mg/d PO given bid-tid; may titrate to max 600 mg/d divided bid-tid. [Trade only: Caps 25, 50, 75, 100, 150, 200, 225, 300 mg.] ▶K ♀C ▶? ©V $$$$$

primidone (***Mysoline***): Start 100-125 mg PO qhs. Increase over 10d to 250 mg tid-qid. Metabolized to phenobarb. [Generic/Trade: Tabs 50,250 mg.] ▶LK ♀D ▶- $$$

tiagabine (***Gabitril***): Start 4 mg PO daily. Increase by 4-8 mg/wk prn to max 32 mg/d (children ≥12 yo) or 56 mg/d (adults) divided bid-qid. [Trade only: Tabs 2, 4, 12, 16 mg.] ▶L ♀C ▶? $$$$$

topiramate (***Topamax***): Partial seizures or primary generalized tonic-clonic seizures, monotherapy (age >10 yrs): Start 25 mg PO bid (week 1), 50 mg bid (week 2), 75 mg bid (week 3), 100 mg bid (week 4), 150 mg bid (week 5), then 200 mg bid as tolerated. Partial seizures, primary generalized tonic-clonic seizures, or Lennox Gastaut Syndrome, adjunctive therapy: Start 25-50 mg PO qhs. Increase weekly by 25-50 mg/d to usual effective dose of 200 mg PO bid. Doses >400 mg/d not shown to be more effective. Migraine prophylaxis: 50 mg PO bid. [Trade only: Tabs 25, 100, 200 mg. Sprinkle Caps 15, 25 mg.] ▶K ♀C ▶? $$$$$

valproic acid (***Depakene, Depakote, Depakote ER, Depacon, divalproex, sodium valproate, ♥Epiject, Epival, Deproic***): Epilepsy: 10-15 mg/kg/d PO/IV divided bid-qid (standard release or IV) or given once daily (Depakote ER). Titrate to max 60 mg/kg/d. Parenteral (Depacon): ≤20 mg/min IV. Hepatotoxicity, drug interactions, reduce dose in elderly. [Generic/Trade: Caps 250mg (Depakene), syrup (Depakene, valproic acid) 250 mg/5 mL. Trade only (Depakote): Caps, sprinkle 125 mg, delayed release tabs 125, 250, 500 mg; extended release tabs (Depakote ER) 250, 500mg.] ▶L ♀D ▶+ $$$$

zonisamide (***Zonegran***): Start 100 mg PO daily. Titrate q2 wks to 300-400 mg PO daily (or divided bid). Max 600 mg/d. Contraindicated in sulfa allergy. [Generic/Trade: Caps 25, 50, 100 mg.] ▶LK ♀C ▶? $$$$

Migraine Therapy - Triptans (5-HT1 Receptor Agonists)

NOTE: May cause vasospasm. Avoid in ischemic or vasospastic heart disease, cerebrovascular syndromes, peripheral arterial disease, uncontrolled HTN, and hemiplegic or basilar migraine. Do not use within 24 hours of ergots or other triptans. Risk of serotonin syndrome with SSRIs, MAOIs.

almotriptan (***Axert***): 6.25-12.5 mg PO. May repeat in 2h prn. Max 25 mg/d. [Trade only: Tabs 6.25, 12.5 mg.] ▶LK ♀C ▶? $

eletriptan (*Relpax*): 20-40 mg PO. May repeat in >2 h prn. Max 40 mg/dose or 80 mg/d. [Trade only: Tabs 20, 40 mg.] ▶LK ♀C ▶? $$

frovatriptan (*Frova*): 2.5 mg PO. May repeat in 2h prn. Max 7.5 mg/24h. [Trade only: Tabs 2.5 mg.] ▶LK ♀C ▶? $

naratriptan (*Amerge*): 1-2.5 mg PO. May repeat in 4h prn. Max 5 mg/24h. [Trade only: Tabs 1, 2.5 mg.] ▶KL ♀C ▶? $$

rizatriptan (*Maxalt, Maxalt MLT*): 5-10 mg PO. May repeat in 2h prn. Max 30 mg/24h. MLT form dissolves on tongue without liquids. [Trade only: Tabs 5, 10 mg. Orally disintegrating tabs (MLT) 5, 10 mg.] ▶LK ♀C ▶? $$

sumatriptan (*Imitrex*): 4-6 mg SC. May repeat in 1h prn. Max 12 mg/24h. Tablets: 25-100 mg PO (50 mg most common). May repeat q2h prn with 25-100 mg doses. Max 200 mg/24h. Intranasal spray: 5-20 mg q2h. Max 40 mg/24h. [Trade only: Tabs 25, 50, 100 mg. Nasal spray 5, 20 mg/ spray. Injection 6 mg/0.5 mL, 4 mg and 6 mg cartridges (STATdose System).] ▶K ♀C ▶+ $$

zolmitriptan (*Zomig, Zomig ZMT*): 1.25-2.5 mg PO q2h. Max 10 mg/24h. Orally disintegrating tabs (ZMT) 2.5 mg PO. May repeat in 2h prn. Max 10 mg/24h. Nasal spray: 5 mg (1 spray) in one nostril. May repeat in 2h. Max 10 mg/24h. [Trade only: Tabs 2.5, 5 mg. Orally disintegrating tabs $$$ (ZMT) 2.5, 5 mg. Nasal spray 5 mg/spray.] ▶L ♀C ▶? $$

DERMATOMES

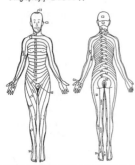

MOTOR NERVE ROOTS

Level	Motor function
C4	Spontaneous breathing
C5	Shoulder shrug / deltoid
C6	Biceps / wrist extension
C7	Triceps / wrist flexion
C8/T1	finger flexion
T1-T12	Intercostal/abd muscles
T12	cremasteric reflex
L1/L2	hip flexion
L2/L3/L4	hip adduction / quads
L5	great toe dorsiflexion
S1/S2	foot plantarflexion
S2-S4	rectal tone

LUMBOSACRAL NERVE ROOT COMPRESSION	Root	Motor	Sensory	Reflex
	L4	quadriceps	medial foot	knee-jerk
	L5	dorsiflexors	dorsum of foot	medial hamstring
	S1	plantarflexors	lateral foot	ankle-jerk

GLASGOW COMA SCALE

Eye Opening	Verbal Activity	Motor Activity
4. Spontaneous	5. Oriented	6. Obeys commands
3. To command	4. Confused	5. Localizes pain
2. To pain	3. Inappropriate	4. Withdraws to pain
1. None	2. Incomprehensible	3. Flexion to pain
	1. None	2. Extension to pain
		1. None

Migraine Therapy - Other

Cafergot (ergotamine + caffeine): 2 tabs (1/100 mg each) PO at onset, then 1 tab q30 min prn. Max 6 tabs/attack or 10/wk. Suppositories (2/100 mg): 1 PR at onset; may repeat in 1h prn. Max 2/attack or 5/wk. [Trade only: Supp 2/100 mg ergotamine/caffeine. Generic: Tabs 1/100 mg ergotamine/caffeine.] ▶L ♀X ▶- $

dihydroergotamine (**D.H.E. 45, Migranal**): Solution (DHE 45) 1 mg IV/IM/SC. May repeat in 1h prn. Max 2 mg (IV) or 3 mg (IM/SC) per day. Nasal spray (Migranal): 1 spray in each nostril. May repeat in 15 min prn. Max 6 sprays/24h or 8 sprays/ wk. [Trade: Nasal spray 0.5 mg/spray. Self-injecting soln: 1 mg/mL.] ▶L ♀X ▶- $$

flunarizine (✚**Sibelium**): Canada only. 10 mg PO qhs. [Generic/Trade: Caps 5 mg] ▶L ♀C ▶- $$

Midrin (isometheptene + dichloralphenazone + acetaminophen, **Amidrine, Durdrin, Migquin, Migratine, Migrazone, Va-Zone**): Tension and vascular headache treatment: 1-2 caps PO q4h. Max 8 caps/d. Migraine treatment: 2 caps PO x 1, then 1 cap q1h prn to max 5 caps/12h. [Generic/Trade: Caps 65/100/325 mg of isometheptene/dichloralphenazone/acetaminophen.] ▶L ♀? ▶? ©IV $

Multiple sclerosis

glatiramer (**Copaxone**): Multiple sclerosis: 20 mg SC daily. [Trade only: Injection 20 mg single dose vial.] ▶Serum ♀B ▶? $$$$$

interferon beta-1A (**Avonex, Rebif**): Multiple sclerosis: Avonex- 30 mcg (6 million units) IM q wk. Rebif- start 8.8 mcg SC three times weekly; and titrate over 4 wks to maintenance dose of 44 mcg three times weekly over 4 wks. Follow LFTs and CBC. [Trade only: Avonex: Injection 33 mcg (6.6 million units) single dose vial. Rebif: Starter kit 20 mcg and 44 mcg pre-filled syringe.] ▶L ♀C ▶? $$$$$

interferon beta-1B (**Betaseron**): Multiple sclerosis: Start 0.0625 mg SC qod; titrate over six weeks to 0.25 mg (8 million units) SC qod. [Trade only: Injection 0.3 mg (9.6 million units) single dose vial.] ▶L ♀C ▶? $$$$$

natalizumab (**Tysabri**): Relapsing multiple sclerosis (monotherapy): 300 mg IV infusion over 1 h q4 wks. ▶Serum ♀C ▶?

Myasthenia Gravis

edrophonium (**Tensilon, Enlon, Reversol**): Evaluation for myasthenia gravis: 2 mg IV over 15-30 seconds (test dose) while on cardiac monitor, then 8 mg IV after 45 sec. Atropine should be readily available in case of cholinergic reaction. Duration of effect is 5-10 min. ▶Plasma ♀C ▶? $

neostigmine (**Prostigmin**): 15-375 mg/d PO in divided doses, or 0.5 mg IM/SC. [Trade only: Tabs 15 mg.] ▶L ♀C ▶? $$$

pyridostigmine (**Mestinon, Mestinon Timespan, Regonal**): Myasthenia gravis: 60-200 mg PO tid (standard release) or 180 mg PO daily or divided bid (extended release). [Trade only: Tabs 60 mg. Extended release tabs 180 mg. Syrup 60 mg/5 mL.] ▶Plasma, K ♀C ▶+ $$

Parkinsonian Agents - Anticholinergics

benztropine mesylate (**Cogentin**): Parkinsonism: 0.5-2 mg IM/PO/IV given once daily or divided bid. EPS: 1-4 mg PO/IM/IV given once daily or divided bid. [Generic/Trade: Tabs 0.5, 1, 2 mg.] ▶LK ♀C ▶? $$

biperiden (**Akineton**): 2 mg PO tid-qid. [Trade only: Tabs 2 mg] ▶LK ♀C ▶? $$

trihexyphenidyl (**Artane**): Start 1 mg PO daily. Increase gradually to 6-10 mg/d, divided tid. Max 15 mg/d. [Generic/Trade: Tabs 2, 5 mg. Elixir 2 mg/5 mL.] ▶LK ♀C ▶? $

Parkinsonian Agents - COMT Inhibitors

entacapone (***Comtan***): Start 200 mg PO with each dose of carbidopa/levodopa. Max 8 tabs/d (1600 mg). [Trade only: Tabs 200 mg.] ▶L ♀C ▶? $$$$

tolcapone (***Tasmar***): Start 100 mg PO tid. Max 600 mg/d. Adjunctive therapy with carbidopa/levodopa. Potentially fatal hepatotoxicity; monitor LFTs. [Trade only: Tabs 100, 200 mg.] ▶LK ♀C ▶? $$$$

Parkinsonian Agents - Dopaminergic Agents & Combinations

amantadine (***Symmetrel***, **♣***Endantadine***): 100 mg PO bid. Max 300-400 mg/d divided tid-qid. [Trade only: Tabs 100 mg. Generic: Caps 100 mg. Generic/Trade: Syrup 50 mg/ 5 mL.] ▶K ♀C ▶- $

apomorphine (***Apokyn***): Start 0.2 ml SC prn. May increase in 0.1 ml increments every few days. Monitor for orthostatic hypotension after initial dose and with dose escalation. Max 0.6 ml/dose or 2 ml/d. Potent emetic - pretreat with trimethobenzamide 300 mg PO tid starting 3d prior to use, and continue for ≥2 months. Contains sulfites. [Trade only: Cartridges (for injector pen, 10 mg/mL) 3 mL. Ampules (10 mg/mL) 2 mL.] ▶L ♀C ▶? $$$$$

bromocriptine (***Parlodel***): Start 1.25 mg PO bid. Usual effective dose 10-40 mg/d. Max 100 mg/d. [Generic/Trade: Tabs 2.5 mg. Caps 5 mg.] ▶L ♀B ▶- $$$$$

carbidopa-levodopa (***Atamet, Sinemet, Sinemet CR, Parcopa***): Start 1 tab (25/100 mg) PO tid. Increase q1-4 d as needed. Sustained release: Start 1 tab (50/200 mg) PO bid; increase q 3d as needed. [Generic/Trade: Tabs (carbidopa/levodopa) 10/100, 25/100, 25/250 mg. Tabs, sustained release (Sinemet CR, carbidopa-levodopa ER) 25/100, 50/200 mg. Trade only: orally disintegrating tablet (Parcopa) 10/100, 25/100, 25/250.] ▶L ♀C ▶- $$$$

pergolide (***Permax***): Start 0.05 mg PO daily. Gradually increase to 1 mg PO tid. Max 5 mg/d. [Generic/Trade: Tabs 0.05, 0.25, 1 mg.] ▶K ♀B ▶? $$$$$

pramipexole (***Mirapex***): Start 0.125 mg PO tid. Gradually increase to 0.5-1.5 mg PO tid. [Trade only: Tabs 0.125, 0.25, 0.5, 1, 1.5 mg.] ▶K ♀C ▶? $$$$$

rasagiline (***Azilect***): Parkinson's disease – monotherapy: 1 mg PO daily. Parkinson's disease – adjunctive: 0.5 mg PO daily. Max 1 mg/d. MAOI diet. [Trade: Tabs 0.5, 1 mg.] ▶L ♀C ▶?

ropinirole (***Requip***): Parkinsonism: Start 0.25 mg PO tid, then gradually increase to 1 mg PO tid. Max 24 mg/d. Restless legs syndrome: Start 0.25 mg PO 1-3 hrs before sleep for 2 days, then increase to 0.5 mg/d days 3-7. Increase by 0.5 mg/d at weekly intervals as tolerated to max dose of 4 mg/d given 1-3 hours before sleep. [Trade only: Tabs 0.25, 0.5, 1, 2, 4, 5 mg.] ▶L ♀C ▶? $$$$

selegiline (***Eldepryl, Zelapar***): 5 mg PO q am and q noon. [Generic/Trade: Caps 5 mg. Tabs 5 mg. Trade only (Zelapar): Oral disintegrating tabs 1.25 mg.] ▶LK ♀C ▶? $$$$

Stalevo (carbidopa + levodopa + entacapone): Parkinson's disease (conversion from carbidopa-levodopa +/- entacapone): Start Stalevo tab with same amount of carbidopa-levodopa, titrate to desired response. May need to reduce levodopa dose if not already taking entacapone. [Trade: Tabs (carbidopa/levodopa/entacapone): 12.5/50/200 mg, 25/100/200 mg, 37.5/150/200 mg.] ▶L ♀C ▶- $$$$$

Other Agents

alteplase (***tpa, t-PA, Activase***, **♣***Activase rt-PA***): Thrombolysis for acute ischemic stroke with symptoms ≤3h: 0.9 mg/kg (max 90 mg); give 10% of total dose as an IV bolus, and the remainder IV over 60 min. Multiple exclusions. ▶L ♀C ▶? $$$$$

botulinum toxin type A (*Botox, Botox Cosmetic*): Dose varies based on indication. [Trade only: 100 unit single-use vials.] ▶Not absorbed ♀C ▶? $$$$$

dexamethasone (*Decadron, ✿Dexasone*): Cerebral edema: Load 10-20 mg IV/IM, then give 4 mg IV/IM q6h or 1-3 mg PO tid. Bacterial meningitis: 0.15 mg/kg IV/IM q6h. [Generic/Trade: Tabs 0.25, 0.5, 0.75, 1.0, 1.5, 2, 4, 6 mg. Elixir/ solution 0.5 mg/5 mL. Trade only: Soln concentrate 0.5 mg/ 0.5 mL (Intensol)] ▶L ♀C ▶- $$$

mannitol (*Osmitrol, Resectisol*): Intracranial HTN: 0.25-2 g/kg IV over 30-60 min. ▶K ♀C ▶? $$

meclizine (*Antivert, Bonine, Medivert, Meclicot, Meni-D, ✿Bonamine*): Motion sickness: 25-50 mg PO 1 hr prior to travel, then 25-50 mg PO daily. Vertigo: 25 mg PO q6h prn. [Rx/OTC/Generic/Trade: Tabs 12.5, 25 mg. Chew tabs 25 mg. Rx/Trade only: Tabs 30, 50 mg. Caps 25 mg.] ▶L ♀B ▶? $

nimodipine (*Nimotop*): Subarachnoid hemorrhage: 60 mg PO q4h x 21 d. [Trade only: Caps 30 mg.] ▶L ♀C ▶- $$$$$

oxybate (*Xyrem, GHB, gamma hydroxybutyrate*): Narcolepsy-associated cataplexy or excessive daytime sleepiness: 2.25 g PO qhs. Repeat in 2.5-4h. May increase by 1.5 g/d at >2 wk intervals to max 9 g/d. From a centralized pharmacy. [Trade only: Solution 180 mL (500 mg/mL) supplied with measuring device and child-proof dosing cups.] ▶L ♀B ▶? ©III $$$$$

riluzole (*Rilutek*): ALS: 50 mg PO q12h. Monitor LFTs. [Trade only: Tabs 50 mg.] ▶LK ♀C ▶- $$$$$

tetrabenazine (*✿Nitoman*): Hyperkinetic movement disorders (Canada only): Start 12.5 mg PO bid-tid. May increase by 12.5 mg/d q3-5 d to usual dose of 25 mg PO tid. Max 200 mg/d. Suicidality. [Trade: Tabs 25 mg.] ▶L ♀? ▶- $$$$

OB/GYN

Contraceptives - Other (Oral contraceptives on table on next page.)

etonogestrel (*Implanon*): Contraception: 1 subdermal implant q3 years. ▶L ♀X ▶+

levonorgestrel (*Plan B*): Emergency contraception: 1 tab PO ASAP but within 72h of intercourse. 2nd tab 12h later. [Trade: Kit contains 2 tabs 0.75 mg.] ▶L ♀X ▶- $

NuvaRing (ethinyl estradiol + etonogestrel): Contraception: 1 ring intravaginally x 3 weeks each month. [Trade only: Flexible intravaginal ring, 15 mcg ethinyl estradiol/0.120 mg etonogestrel/day. 1 and 3 rings/box.] ▶L ♀X ▶- $$

Ortho Evra (norelgestromin + ethinyl estradiol, ✿Evra): Contraception: 1 patch q week x 3 weeks, then 1 week patch-free. [Trade only: Transdermal patch: 150/20 mcg norelgestromin/ethinyl estradiol per day. 1 and 3 patches/box.] ▶L ♀X ▶- $$$

Estrogens (See also Hormone Replacement Combinations.)

esterified estrogens (*Menest*): HRT: 0.3 to 1.25 mg PO daily. [Trade only: Tabs 0.3, 0.625, 1.25, 2.5 mg.] ▶L ♀X ▶- $

estradiol (*Estrace, Estradiol, Gynodiol*): HRT: 1-2 mg PO daily. [Generic/Trade: Tabs, micronized 0.5, 1, 2 mg, scored. Tabs: 1.5 mg (Gynodiol).] ▶L ♀X ▶- $

estradiol acetate (*Femtrace*): HRT: 0.45-1.8 mg PO daily. [Trade only: Tabs, 0.45, 0.9, 1.8 mg.] ▶L ♀X ▶- $$

estradiol acetate vaginal ring (*Femring*): Menopausal atrophic vaginitis & vasomotor symptoms: Insert & replace after 90 days. [Trade only: 0.05 mg/day and 0.1 mg/day.] ▶L ♀X ▶- $$

estradiol cypionate (*Depo-Estradiol*): HRT: 1-5 mg IM q 3-4 weeks. ▶L ♀X ▶- $

estradiol gel (*Estrogel*): HRT: Thinly apply contents of one complete pump depres-

ORAL CONTRACEPTIVES* ▶L ♀X Monophasic	Estrogen (mcg)	Progestin (mg)
Norinyl 1+50, Ortho-Novum 1/50, Necon 1/50	50 mestranol	1 norethindrone
Ovcon-50	50 ethinyl estradiol	
Demulen 1/50, Zovia 1/50E		1 ethynodiol
Ovral, Ogestrel		0.5 norgestrel
Norinyl 1+35, Ortho-Novum 1/35, Necon 1/35, Nortrel 1/35	35 ethinyl estradiol	1 norethindrone
Brevicon, Modicon, Necon 0.5/35, Nortrel 0.5/35		0.5 norethindrone
Ovcon-35		0.4 norethindrone
Previfem		0.18 norgestimate
Ortho-Cyclen, MonoNessa, Sprintec-28		0.25 norgestimate
Demulen 1/35, Zovia 1/35E, Kelnor 1/35		1 ethynodiol
Loestrin 21 1.5/30, Loestrin Fe 1.5/30, Junel 1.5/30, Junel 1.5/30 Fe, Microgestin Fe 1.5/30	30 ethinyl estradiol	1.5 norethindrone
Cryselle, Lo/Ovral, Low-Ogestrel		0.3 norgestrel
Apri, Desogen, Ortho-Cept, Reclipsen		0.15 desogestrel
Levlen, Levora, Nordette, Portia, Seasonale, Seasonique▲		0.15 levonorgestrel
Yasmin		3 drospirenone
Loestrin 21 1/20, Loestrin Fe 1/20, Loestin 24 Fe, Junel 1/20, Junel Fe 1/20, Microgestin Fe 1/20	20 ethinyl estradiol	1 norethindrone
Alesse, Aviane, Lessina, Levlite, Lutera		0.1 levonorgestrel
Yaz		3 drospirenone
Progestin-only		
Micronor, Nor-Q.D., Camila, Errin, Jolivette, Nora-BE	none	0.35 norethindrone
Ovrette		0.075 norgestrel
Biphasic (estrogen & progestin contents vary)		
Kariva, Mircette	20/10 eth estrad	0.15/0 desogestrel
Ortho Novum 10/11, Necon 10/11	35 eth estradiol	0.5/1 norethindrone
Triphasic (estrogen & progestin contents vary)		
Cyclessa, Velivet	25 ethinyl estradiol	0.100/0.125/0.150 desogestrel
Ortho-Novum 7/7/7, Necon 7/7/7, Nortrel 7/7/7	35 ethinyl estradiol	0.5/0.75/1 norethindr
Tri-Norinyl, Leena		0.5/1/0.5 norethindr
Enpresse, Tri-Levlen, Triphasil, Trivora-28	30/40/30 ethinyl estradiol	0.5/0.75/0.125 levonorgestrel
Ortho Tri-Cyclen, Trinessa, Tri-Sprintec, Tri-Previfem	35 eth estradiol	0.18/0.215/0.25 norgestimate
Ortho Tri-Cyclen Lo	25 eth estradiol	
Estrostep Fe	20/30/35 eth estr	1 norethindrone

▲84 light blue-green active pills followed by 7 yellow pills with 10 mg ethinyl estradiol.

*All: Not recommended in smokers. Increase risk of thromboembolism, stroke, MI, hepatic neoplasia & gallbladder disease. Nausea, breast tenderness, & breakthrough bleeding are common transient side effects. Effectiveness reduced by hepatic enzyme-inducing drugs such as certain anticonvulsants and barbiturates, rifampin, rifabutin, griseofulvin, & protease inhibitors. Coadministration with St. John's wort may decrease efficacy. Vomiting or diarrhea may also increase the risk of contraceptive failure. Consider an additional form of birth control in above circumstances. See product insert for instructions on missing doses. Most available in 21 and 28 day packs. **Progestin only**: Must be taken at the same time every day. Because much of the literature regarding OC adverse effects pertains mainly to estrogen/progestin combinations, the extent to which progestin-

sion (1.25 g) to one entire arm. [Trade only: Gel 0.06% in non-aerosol, metered-dose pump with 64 1.25 g doses.] ▶L ♀X ▶- $$

estradiol topical emulsion (*Estrasorb*): HRT: Rub in contents of one pouch each to left and right legs (spread over thighs & calves) qam. Daily dose = two 1.74 g pouches. [Trade only: Topical emulsion, 56 pouches/carton.] ▶L ♀X ▶- $$$

estradiol transdermal system (*Alora, Climara, Esclim, Estraderm, FemPatch, Menostar, Vivelle, Vivelle Dot, ♣Estradot, Oesclim*): HRT: Apply one patch weekly (Climara, FemPatch, Estradiol, Menostar) or twice per week (Esclim, Estraderm, Vivelle, Vivelle Dot, Alora). [Generic/Trade: Transdermal patches doses in mg/day: Climara (q week) 0.025, 0.0375, 0.05, 0.06, 0.075, 0.1. Trade only: FemPatch (q week) 0.025. Esclim (twice/week) 0.025, 0.0375, 0.05, 0.075, 0.1. Vivelle, Vivelle Dot (twice/week) 0.025, 0.0375, 0.05, 0.075, 0.1. Estraderm (twice/week) 0.05, & 0.1. Alora (twice/week) 0.025, 0.05, 0.075, 0.1.] ▶L ♀X ▶- $$$

estradiol vaginal ring (*Estring*): Menopausal atrophic vaginitis: Insert & replace after 90 days. [Trade only: 2 mg ring single pack.] ▶L ♀X ▶- $$$

estradiol vaginal tab (*Vagifem*): Menopausal atrophic vaginitis: one tablet vaginally daily x 2 weeks, then one tablet vaginally 2x/week. [Trade only: Vaginal tab: 25 mcg in disposable single-use applicators, 8 & 18/pack.] ▶L ♀X ▶- $$$

estradiol valerate (*Delestrogen*): HRT: 10-20 mg IM q4 weeks. ▶L ♀X ▶- $

estrogen cream (*Premarin, Estrace*): Menopausal atrophic vaginitis: Premarin: 0.5-2 g daily. Estrace: 2-4 g daily x 2 weeks, then reduce. [Trade only: Vaginal cream. Premarin: 0.625 mg conjugated estrogens/g in 42.5 g with or w/o calibrated applicator. Estrace: 0.1 mg estradiol/g in 42.5 g w/calibrated applicator.] ▶L ♀X ▶? $$$

estrogens conjugated (*Premarin, C.E.S., Congest*): HRT: 0.3 to 1.25 mg PO daily. Abnormal uterine bleeding: 25 mg IV/IM. Repeat in 6-12h if needed. [Trade only: Tabs 0.3, 0.45, 0.625, 0.9, 1.25 mg.] ▶L ♀X ▶- $$

estrogens synthetic conjugated A (*Cenestin*): HRT: 0.3 to 1.25 mg PO daily. [Trade only: Tabs 0.3, 0.45, 0.625, 0.9, 1.25 mg.] ▶L ♀X ▶- $$

estrogens synthetic conjugated B (*Enjuvia*): HRT: 0.3 to 1.25 mg PO daily. [Trade only: Tabs 0.3, 0.45, 0.625, 1.25 mg.] ▶L ♀X ▶- $$

estropipate (*Ogen, Ortho-Est*): HRT: 0.75 to 6 mg PO daily. [Generic/Trade: Tabs 0.75, 1.5, 3, 6 mg of estropipate.] ▶L ♀X ▶- $

GnRH Agents

cetrorelix acetate (*Cetrotide*): Infertility: 0.25 mg SC daily during the early to mid follicular phase or 3 mg SC x1 usually on stimulation day 7. ▶Plasma ♀X ▶- $$$$$

ganirelix (*Follistim-Antagon Kit, ♣Orgalutran*): Infertility: 250 mcg SC daily during the early to mid follicular phase. ▶Plasma ♀X ▶? $$$$$

goserelin (*Zoladex*): Endometriosis: 3.6 mg implant SC q28 days or 10.8 mg implant q12 weeks x 6 months. ▶LK ♀X ▶- $$$$$

leuprolide (*Lupron, Lupron Depot*): Endometriosis or fibroid-associated anemia, Depot: 3.75 mg IM q month or 11.25 mg IM q3 months, for total therapy of 6 months (endometriosis) or 3 months (fibroids). ▶L ♀X ▶- $$$$$

nafarelin (*Synarel*): Endometriosis: 200-400 mcg intranasal bid x 6 months. [Trade: Nasal soln 2 mg/mL (200 mcg/spray) about 80 sprays/bottle.] ▶L ♀X ▶- $$$$$

ORAL CONTRACEPTIVES, continued footnotes

only contraceptives cause these effects is unclear. No significant interaction has been found with broad-spectrum antibiotics. The effect of St. John's wort is unclear. No placebo days, start new pack immediately after finishing current one. Available in 28 day packs. Readers may find the following website useful: www.managingcontraception.com.

EMERGENCY CONTRACEPTION within 72 hours of unprotected sex: Take first dose ASAP, then identical dose 12h later. *Plan B* kit contains 2 levonorgestrel 0.75mg tabs. Each dose is 1 pill. The progestin-only method causes less nausea & may be more effective. Alternate regimens: Each dose is either 2 pills of *Ovral* or *Ogestrel*, 4 pills of *Cryselle, Levlen, Levora, Lo/Ovral, Nordette, Tri-Levlen*, Tri-phasil*, Trivora*, or Low Ogestrel*, or 5 pills of *Alesse, Aviane, Lessina, or Levlite*. If vomiting occurs within 1 hour of taking either dose of medication, consider whether or not to repeat that dose with an antiemetic 1h prior. More info at: www.not-2-late.com.   *Use 0.125 mg levonorgestrel/30 mcg ethinyl estradiol tabs.

Hormone Replacement Combinations (See also estrogens)

Activella (estradiol + norethindrone): HRT: 1 tab PO daily. [Trade: Tab 1 mg estradiol/0.5 mg norethindrone acetate in calendar dial pack dispenser.] ▶L ♀X ▶- $$

Angeliq (estradiol + drospirenone): HRT: 1 tab PO daily. [Trade only: Tabs 1 mg estradiol/0.5 mg drospirenone.] ▶L ♀X ▶- ?

Climara Pro (estradiol + levonorgestrel): HRT: 1 patch weekly. [Trade: Transdermal 0.045/0.015 estradiol/levonorgestrel in mg/day, 4 patches/box.] ▶L ♀X ▶- $$

CombiPatch (estradiol + norethindrone, ✿*Estalis*): HRT: 1 patch twice weekly. [Trade only: Transdermal patch 0.05 estradiol/ 0.14 norethindrone & 0.05 estradiol/0.25 norethindrone in mg/day, 8 patches/box.] ▶L ♀X ▶- $$

Estratest (esterified estrogens + methyltestosterone): HRT: 1 tab PO daily. [Trade: Tabs 1.25 mg esterified estrogens/2.5 mg methyltestosterone.] ▶L ♀X ▶- $$$

Estratest H.S. (esterified estrogens + methyltestosterone): HRT: 1 tab PO daily. [Trade: Tabs 0.625 mg esterified estrogens/1.25 mg methyltestosterone.] ▶L ♀X ▶- $$$

FemHRT (ethinyl estradiol + norethindrone): HRT: 1 tab PO daily. [Trade: Tabs 5/1 and 2.5/0.5 mcg ethinyl estradiol/mg norethindrone, 28/blister card.] ▶L ♀X ▶- $$

Prefest (estradiol + norgestimate): HRT: 1 pink tab PO daily x 3 days followed by 1 white tab PO daily x 3 days, sequentially throughout the month. [Trade only: Tabs in 30-day blister packs 1 mg estradiol (15 pink) & 1 mg estadiol/0.09 mg norgestimate (15 white).] ▶L ♀X ▶- $$

Premphase (estrogens conjugated + medroxyprogesterone): HRT: 1 tab PO daily. [Trade: Tabs in 28-day EZ-Dial dispensers: 0.625 mg conjugated estrogens (14) & 0.625 mg/5 mg conjugated estrogens/medroxyprogesterone (14).] ▶L ♀X ▶- $$

Prempro (estrogens conjugated + medroxyprogesterone, ✿*PremPlus*): HRT: 1 tab PO daily. [Trade only: Tabs in 28-day EZ-Dial dispensers: 0.625 mg/5 mg, 0.625 mg/2.5 mg, 0.45 mg/1.5 mg, or 0.3 mg/1.5 mg conjugated estrogens/medroxyprogesterone.] ▶L ♀X ▶- $$

Syntest D.S. (esterified estrogens + methyltestosterone): HRT: 1 tab PO daily. [Trade: Tabs 1.25/2.5 mg esterified estrogens/methyltestosterone.] ▶L ♀X ▶- $$

Syntest H.S. (esterified estrogens + methyltestosterone): HRT: 1 tab PO daily. [Trade: Tabs 0.625/1.25 mg esterified estrogens/methyltestosterone.] ▶L ♀X ▶- $$

Labor Induction / Cervical Ripening

dinoprostone (*PGE2, Prepidil, Cervidil, Prostin E2*): Cervical ripening: One syringe of gel placed directly into the cervical os for cervical ripening or one insert in the posterior fornix of the vagina. [Trade: Gel (Prepidil) 0.5 mg/3 g syringe. Vaginal insert (Cervidil) 10 mg. Vag supp (Prostin E2) 20 mg.] ▶Lung ♀C ▶? $$$$$

misoprostol (*PGE1, Cytotec*): Cervical ripening: 25 mcg intravaginally q3-6h (or 50 mcg q6h). First-trimester pregnancy failure: 800 mcg intravaginally, repeat on day 3 if expulsion incomplete. [Generic/Trade: Oral tabs 100 & 200 mcg.] ▶LK ♀X ▶- $

oxytocin (**Pitocin**): Labor induction: 10 units in 1000 mL NS (10 milliunits/mL), start at 6-12 mL/h (1-2 milliunits/min). Uterine contractions/postpartum bleeding: 10 units IM or 10-40 units in 1000 mL NS IV, infuse 20-40 milliunits/min. ▶LK ♀? ▶- $

Ovulation Stimulants

clomiphene (**Clomid, Serophene**): Specialized dosing for ovulation induction. [Generic/Trade: Tabs 50 mg, scored.] ▶L ♀D ▶? $$

Progestins

hydroxyprogesterone caproate: Amenorrhea, dysfunctional uterine bleeding, metrorrhagia: 375 mg IM. Production of secretory endometrium & desquamation: 125-250 mg IM on 10th day of the cycle, repeat q7days until suppression no longer desired. ▶L ♀X ▶? $$

medroxyprogesterone (**Provera, Amen**): HRT: 10 mg PO daily for last 10-12 days of month, or 2.5-5 mg PO daily. Secondary amenorrhea, abnormal uterine bleeding: 5-10 mg PO daily x 5-10 days. Endometrial hyperplasia: 10-30 mg PO daily. [Generic/Trade: Tabs 2.5, 5, & 10 mg, scored.] ▶L ♀X ▶+ $

medroxyprogesterone - injectable (**Depo-Provera, depo-subQ provera 104**): Contraception/Endometriosis: 150 mg IM in deltoid or gluteus maximus or 104 mg SC in anterior thigh or abdomen q13 weeks. ▶L ♀X ▶+ $$

megestrol (**Megace, Megace ES**): Endometrial hyperplasia: 40-160 mg PO daily x 3-4 mo. AIDS anorexia: 800 mg (20 mL) susp PO daily or 625 mg (5 mL) ES daily. [Generic/Trade: Tabs 20 & 40 mg. Suspension 40 mg/mL in 240 mL. Trade only: Megace ES suspension 125 mg/mL (150 mL).] ▶L ♀D ▶? $$$$$

norethindrone (**Aygestin**): Amenorrhea, abnormal uterine bleeding: 2.5-10 mg PO daily x 5-10 days during the second half of the menstrual cycle. Endometriosis: 5 mg PO daily x 2 weeks. Increase by 2.5 mg q 2 weeks to 15 mg. [Generic/Trade: Tabs 5 mg, scored.] ▶L ♀D ▶? $

progesterone gel (**Crinone, Prochieve**): Secondary amenorrhea: 45 mg (4%) intravaginally qod up to 6 doses. If no response, use 90 mg (8%) qod up to 6 doses. Infertility: special dosing. [Trade only: 4%, 8% single-use, prefilled applicators.] ▶Plasma ♀- ▶? $$$

progesterone micronized (**Prometrium**): HRT: 200 mg PO qhs 10-12 days/month or 100 mg qhs daily. Secondary amenorrhea: 400 mg PO qhs x 10 days. Contraindicated in peanut allergy. [Trade only: Caps 100 & 200 mg.] ▶L ♀B ▶+ $$

DRUGS GENERALLY ACCEPTED AS SAFE IN PREGNANCY (selected)

<u>Analgesics:</u> acetaminophen, codeine*, meperidine*, methadone*. <u>Antimicrobials:</u> penicillins, cephalosporins, erythromycins (not estolate), azithromycin, nystatin, clotrimazole, metronidazole**, nitrofurantoin***, Nix. Antivirals: acyclovir, valacyclovir, famciclovir. <u>CV:</u> labetalol, methyldopa, hydralazine. <u>Derm:</u> erythromycin, clindamycin, benzoyl peroxide. <u>Endo:</u> insulin, liothyronine, levothyroxine. <u>ENT:</u> chlorpheniramine, diphenhydramine, dimenhydrinate, dextromethorphan, guaifenesin, nasal steroids, nasal cromolyn. <u>GI:</u> trimethobenzamide, antacids*, simethicone, cimetidine, famotidine, ranitidine, nizatidine, psyllium, metoclopramide, bisacodyl, docusate, doxylamine, meclizine. <u>Psych:</u> fluoxetine****, desipramine, doxepin. <u>Pulmonary:</u> short-acting inhaled beta-2 agonists, cromolyn, nedocromil, beclomethasone, budesonide, theophylline, prednisone**. <u>Other</u> - heparin.
*Except if used long-term or in high does at term **Except 1st trimester.
***Contraindicated at term and during labor and delivery. ****Except 3rd trimester.

APGAR SCORE	Heart rate	0. Absent	1. <100	2. >100
	Respirations	0. Absent	1. Slow/irreg	2. Good/crying
	Muscle tone	0. Limp	1. Some flexion	2. Active motion
	Reflex irritability	0. No response	1. Grimace	2. Cough/sneeze
	Color	0. Blue	1. Blue extremities	2. Pink

Selective Estrogen Receptor Modulators

raloxifene (**Evista**): Osteoporosis prevention/treatment: 60 mg PO daily. [Trade: Tabs 60 mg.] ▶L ♀X ▶- $$$

tamoxifen (**Nolvadex, Soltamox, Tamone, ♣Tamofen**): Breast cancer prevention: 20 mg PO daily x 5 years. Breast cancer: 10-20 mg PO bid. [Generic/Trade: Tabs 10 & 20 mg, Trade only (Soltamox): soln 10mg/5mL (150 mL).] ▶L ♀D ▶- $$$

Tocolytics

indomethacin (**Indocin, Indocid**): Preterm labor: initial 50-100 mg PO followed by 25 mg PO q6-12h up to 48 hrs. [Generic/Trade: Immediate-release caps 25 & 50 mg. Trade: Oral suspension 25 mg/5 mL.] ▶L ♀? ▶- $

magnesium sulfate: Eclampsia: 4-6 g IV over 30 min, then 1-2 g/h. Drip: 5 g in 250 mL D5W (20 mg/mL), 2 g/h = 100 mL/h. Preterm labor: 6 g IV over 20 minutes, then 2-3 g/h titrated to decrease contractions. Monitor respirations & reflexes. If needed, may reverse toxic effects with calcium gluconate 1g IV. ▶K ♀A ▶+ $$

nifedipine (**Procardia, Adalat, Procardia XL, Adalat CC, ♣Adalat XL**): Preterm labor: loading dose: 10 mg PO q20-30 min if contractions persist, up to 40 mg within the first hour. Maintenance dose: 10-20 mg PO q4-6h or 60-160 mg extended release PO daily. [Generic/Trade: immediate release caps 10 & 20 mg. Extended release tabs 30, 60, 90 mg.] ▶L ♀C ▶+ $

terbutaline (**Brethine, Bricanyl**): Preterm labor: 0.25 mg SC q30 min up to 1 mg/4 hours. Infusion: 2.5-10 mcg/min IV, gradually increased to effective max doses of 17.5-30 mcg/minute. Tachycardia common. [Generic/Trade: Tabs 2.5, 5 mg (Brethine scored)] ▶L ♀B ▶+ $$$$$

Uterotonics

carboprost (**Hemabate, 15-methyl-prostaglandin F2 alpha**): Refractory postpartum uterine bleeding: 250 mcg deep IM. ▶LK ♀C ▶? $$$

methylergonovine (**Methergine**): Refractory postpartum uterine bleeding: 0.2 mg IM/PO tid-qid prn. [Trade only: Tabs 0.2 mg.] ▶LK ♀C ▶? $$

Vaginitis Preparations (See also STD/vaginitis table in antimicrobial section.)

boric acid: Resistant vulvovaginal candidiasis: 1 vag supp qhs x 2 weeks. [No commercial preparation; must be compounded by pharmacist. Vaginal suppositories 600 mg in gelatin capsules.] ▶Not absorbed ♀? ▶- $

butoconazole (**Gynazole, Mycelex-3**): Vulvovaginal candidiasis: Mycelex 3: 1 applicatorful qhs x 3-6 days. Gynazole-1: 1 applicatorful intravaginally qhs x 1. [OTC: Trade only (Mycelex 3): 2% vaginal cream in 5 g pre-filled applicators (3s) & 20 g tube with applicators. Rx: Trade only (Gynazole-1): 2% vaginal cream in 5 g pre-filled applicator.] ▶LK ♀C ▶? $(OTC)

clindamycin (**Cleocin, Clindesse, ♣Dalacin**): Bacterial vaginosis: Cleocin: 1 applicatorful cream qhs x 7d or one vaginal supp qhs x 3d. Clindesse: 1 applicatorful cream x 1. [Generic/Trade: 2% vaginal cream in 40 g tube with 7 disposable applicators (Cleocin). Vag supp (Cleocin Ovules) 100 mg (3) w/applicator. 2% vaginal cream in a single-dose prefilled applicator (Clindesse).] ▶L ♀- ▶+ $$

clotrimazole (**Mycelex 7, Gyne-Lotrimin**, ✤**Canesten, Clotrimaderm**): Vulvovaginal candidiasis: 1 applicatorful 1% cream qhs x 7 days. 1 applicatorful 2% cream qhs x 3 days. 1 vag supp 100 mg qhs x 7 days. 1 vag tab 200 mg qhs x 3 days. [OTC: Generic/Trade: 1% vaginal cream with applicator (some pre-filled). 2% vaginal cream with applicator. Vaginal suppositories 100 mg (7) & 200 mg (3) with applicators. 1% topical cream in some combination packs.] ▶LK ♀B ▶? $

metronidazole - vaginal (**MetroGel-Vaginal, Vandazole**): Bacterial vaginosis: 1 applicatorful qhs or bid x 5 days. [Trade only: 0.75% gel in 70 g tube with applicator.] ▶LK ♀B ▶? $$$

miconazole (**Monistat, Femizol-M, M-Zole, Micozole, Monazole**): Vulvovaginal candidiasis: 1 applicatorful qhs x 3 (4%) or 7 (2%) days. 100 mg vag supp qhs x 7 days. 400 mg vag supp qhs x 3 days. 1200 mg vag supp x 1. [OTC: Generic/Trade: 2% vaginal cream in 45 g with 1 applicator or 7 disposable applicators. Vaginal suppositories 100 mg (7) OTC: Trade: 400 mg (3) & 1200 mg (1) with applicator. Generic/Trade: 4% vaginal cream in 25 g tubes or 3 prefilled applicators. Some in combination packs with 2% miconazole cream for external use.] ▶LK ♀+ ▶? $

nystatin (**Mycostatin**, ✤**Nilstat, Nyaderm**): Vulvovaginal candidiasis: 1 vag tab qhs x 14 days. [Generic/Trade: Vaginal Tabs 100,000 units in 15s & 30s with or without applicator(s).] ▶Not metabolized ♀A ▶? $$

terconazole (**Terazol**): Vulvovaginal candidiasis: 1 applicatorful of 0.4% cream qhs x 7 days, or 1 applicatorful of 0.8% cream qhs x 3 days, or 80 mg vag supp qhs x 3 days. [All forms supplied with applicators: Generic/Trade: Vag cream 0.4% (Terazol 7) in 45 g tube, 0.8% (Terazol 3) in 20 g tube. Vag supp (Terazol 3) 80 mg (#3).] ▶LK ♀C ▶- $$

tioconazole (**Monistat 1-Day, Vagistat-1**): Vulvovaginal candidiasis: 1 applicatorful of 6.5% ointment intravaginally qhs single-dose. [OTC: Trade only: Vaginal ointment: 6.5% (300 mg) in 4.6 g prefilled single-dose applicator.] ▶Not absorbed ♀C ▶- $

Other OB/GYN Agents

clonidine (**Catapres, Catapres-TTS**, ✤**Dixarit**): Menopausal flushing: 0.1-0.4 mg/day PO divided bid-tid. Transdermal system applied weekly: 0.1 mg/day. [Generic/Trade: Tabs, non-scored 0.1, 0.2, 0.3 mg. Trade only: transdermal weekly patch 0.1 mg/day (TTS-1), 0.2 mg/day (TTS-2), 0.3 mg/day (TTS-3).] ▶LK ♀C ▶? $

danazol (**Danocrine**, ✤**Cyclomen**): Endometriosis: Start 400 mg PO bid, then titrate downward to maintain amenorrhea x 3-6 months. Fibrocystic breast disease: 100-200 mg PO bid x 4-6 months. [Generic/Trade: Caps 50, 100, 200 mg.] ▶L ♀X ▶- $$$$

mifepristone (**Mifeprex, RU-486**): 600 mg PO x 1 followed by 400 mcg misoprostol on day 3, if abortion not confirmed. [Trade only: Tabs 200 mg.] ▶L ♀X ▶? $$$$$

Premesis-Rx (B6 + folic acid + B12 + calcium carbonate): Pregnancy-induced nausea: 1 tab PO daily. [Trade only: Tabs 75 mg vitamin B6 (pyridoxine), sustained-release, 12 mcg vitamin B12 (cyanocobalamin), 1 mg folic acid, and 200 mg calcium carbonate.] ▶L ♀A ▶+ $

RHO immune globulin (**RhoGAM, BayRho-D, WinRho SDF, MICRhoGAM, BayRho-D Mini Dose**): 300 mcg vial IM to mother at 28 weeks gestation followed by a 2nd dose ≤72 hours of delivery (if mother Rh- and baby is or might be Rh+). Microdose (50 mcg, MICRhoGAM) OK if spontaneous abortion <12 weeks gestation. ▶L ♀C ▶? $$$$$

ONCOLOGY

Alkylating agents: altretamine (*Hexalen*), busulfan (*Myleran, Busulfex*), carmustine (*BCNU, BiCNU, Gliadel*), chlorambucil (*Leukeran*), cyclophosphamide (*Cytoxan, Neosar*), dacarbazine (*DTIC-Dome*), ifosfamide (*Ifex*), lomustine (*CeeNu, CCNU*), mechlorethamine (*Mustargen*), melphalan (*Alkeran*), procarbazine (*Matulane*), streptozocin (*Zanosar*), temozolomide (*Temodar*, ❤*Temodal*), thiotepa (*Thioplex*). **Antibiotics:** bleomycin (*Blenoxane*), dactinomycin (*Cosmegen*), daunorubicin (*DaunoXome, Cerubidine*), doxorubicin liposomal (*Doxil*, ❤*Caelyx, Myocet*), doxorubicin non-liposomal (*Adriamycin, Rubex*), epirubicin (*Ellence*, ❤*Pharmorubicin*), idarubicin (*Idamycin*), mitomycin (*Mutamycin, Mitomycin-C*), mitoxantrone (*Novantrone*), valrubicin (*Valstar*, ❤*Valtaxin*). **Antimetabolites:** azacitidine (*Vidaza*), capecitabine (*Xeloda*), cladribine (*Leustatin, chlorodeoxyadenosine*), clofarabine (*Clolar*), cytarabine (*Cytosar-U, Tarabine, Depo-Cyt, AraC*), decitabine (*Dacogen*), floxuridine (*FUDR*), fludarabine (*Fludara*), fluorouracil (*Adrucil, 5-FU*), gemcitabine (*Gemzar*), hydroxyurea (*Hydrea, Droxia*), mercaptopurine (*6-MP, Purinethol*), methotrexate (*Rheumatrex, Trexall*), nelarabine (*Arranon*), pemetrexed (*Alimta*), pentostatin (*Nipent*), thioguanine (*Tabloid*, ❤*Lanvis*). **Cytoprotective Agents:** amifostine (*Ethyol*), dexrazoxane (*Zinecard*), mesna (*Mesnex*, ❤*Uromitexan*), palifermin (*Kepivance*). **Hormones:** abarelix (*Plenaxis*), anastrozole (*Arimidex*), bicalutamide (*Casodex*), cyproterone (❤*Androcur, Androcur Depot*), estramustine (*Emcyt*), exemestane (*Aromasin*), flutamide (*Eulexin*, ❤*Euflex*), fulvestrant (*Faslodex*), goserelin (*Zoladex*), histrelin (*Vantas*), letrozole (*Femara*), leuprolide (*Eligard, Lupron, Oaklide, Viadur*), nilutamide (*Nilandron*, ❤*Anandron*), testolactone (*Teslac*), toremifene (*Fareston*), triptorelin (*Trelstar Depot*). **Immunomodulators:** aldesleukin (*Proleukin, interleukin-2*), alemtuzumab (*Campath*), BCG (*Bacillus of Calmette & Guerin, Pacis, TheraCys, TICE BCG*, ❤*Oncotice, Immucyst, Pacis*), bevacizumab (*Avastin*), cetuximab (*Erbitux*), dasatinib (*Sprycel*), denileukin (*Ontak*), erlotinib (*Tarceva*), gemtuzumab (*Mylotarg*), ibritumomab (*Zevalin*), imatinib (*Gleevec*), interferon alfa-2a (*Roferon-A*), interferon alfa-2b (*Intron A*), interferon alfa-n3 (*Alferon N*), rituximab (*Rituxan*), sunitinib (*Sutent*), tositumomab (*Bexxar*), trastuzumab (*Herceptin*). **Mitotic Inhibitors:** docetaxel (*Taxotere*), etoposide (*VP-16, Etopophos, Toposar, VePesid*), paclitaxel (*Taxol, Abraxane, Onxol*), teniposide (*Vumon, VM-26*), vinblastine (*Velban, VLB*), vincristine (*Oncovin, Vincasar, VCR*), vinorelbine (*Navelbine*). **Platinum-Containing Agents:** carboplatin (*Paraplatin*), cisplatin (*Platinol-AQ*), oxaliplatin (*Eloxatin*). **Radiopharmaceuticals:** samarium 153 (*Quadramet*), strontium-89 (*Metastron*). **Miscellaneous:** arsenic trioxide (*Trisenox*), asparaginase (*Elspar*, ❤*Kidrolase*), bexarotene (*Targretin*), bortezomib (*Velcade*), gefitinib (*Iressa*), irinotecan (*Camptosar*), lenalidomide (*Revlimid*), leucovorin (*Wellcovorin, folinic acid*), mitotane (*Lysodren*), pegaspargase (*Oncaspar*), porfimer (*Photofrin*), sorafenib (*Nexavar*), topotecan (*Hycamtin*), tretinoin (*Vesanoid*).

OPHTHALMOLOGY

> **NOTE:** Most eye medications can be administered 1 drop at a time despite common manufacturer recommendations of 1-2 drops concurrently. Even a single drop is typically more than the eye can hold and thus a second drop is both wasteful and increases the possibility of systemic toxicity. If twice the medication is desired separate single drops by at least 5 minutes.

Antiallergy - Decongestants & Combinations
naphazoline (*Albalon, AK-Con, Vasocon, Naphcon, Allerest, Clear Eyes*): 1 gtt qid prn for up to 4 days. [OTC Generic/Trade: solution 0.012, 0.02, 0.03% (15, 30 mL). Rx Generic/Trade: 0.1% (15 mL).] ▶? ♀C ▶? $

Naphcon-A (naphazoline + pheniramine, Visine-A): 1 gtt qid prn for up to 4 days. [OTC Trade only: solution 0.025% + 0.3% (15 mL).] ▶L ♀C ▶? $

Vascon-A (naphazoline + antazoline): 1 gtt qid prn for up to 4 days. [OTC Trade only: solution 0.1% + 0.5% (15 mL).] ▶L ♀C ▶? $

Antiallergy - Dual Antihistamine & Mast Cell Stabilizer

azelastine (***Optivar***): 1 gtt bid. [Trade only: solution 0.05% (3, 6 mL).] ▶L ♀C ▶? $$

epinastine (***Elestat***): 1 gtt bid. [Trade: solution 0.05% (5,10 mL).] ▶K ♀C ▶? $$$

ketotifen (***Zaditor***): 1 gtt in each eye q8-12h. [Generic/Trade: solution 0.025% (5 mL).] ▶Minimal absorption ♀C ▶? $$$

olopatadine (***Patanol***): 1 gtt in each eye twice daily at an interval of 6-8h (Patanol). [Trade only: solution 0.1% (5 mL).] ▶K ♀C ▶? $$$

Antiallergy - Pure Antihistamines

emedastine (***Emadine***): 1 gtt up to qid. [Trade: soln 0.05% (5 mL).] ▶L ♀B ▶? $$

levocabastine (***Livostin***): 1 gtt qid. [Trade only: suspension 0.05% (5,10 mL).] ▶Minimal absorption ♀C ▶? $$$

Antiallergy - Pure Mast Cell Stabilizers

cromolyn sodium (***Crolom, Opticrom***): 1-2 gtts in each eye 4-6 times per day. [Generic/Trade: solution 4% (10 mL).] ▶LK ♀B ▶? $$$

lodoxamide (***Alomide***): 1-2 gtts in each eye qid. [Trade only: solution 0.1% (10 mL).] ▶K ♀B ▶? $$$

nedocromil - ophthalmic (***Alocril***): 1-2 gtts in each eye bid. [Trade only: solution 2% (5 mL).] ▶L ♀B ▶? $$

pemirolast (***Alamast***): 1-2 gtts in each eye qid. [Trade only: solution 0.1% (10 mL).] ▶? ♀C ▶? $$$

Antibacterials - Aminoglycosides

gentamicin - ophthalmic (***Garamycin, Genoptic, Gentak, ✱Alcomicin, Diogent***): 1-2 gtts q2-4h; ½ inch ribbon of oint bid-tid. [Generic/Trade: solution 0.3% (5 ,15 mL), ointment 0.3% (3.5 g tube).] ▶K ♀C ▶? $

tobramycin - ophthalmic (***Tobrex***): 1-2 gtts q1-4h or ½ inch ribbon of ointment q3-4h or bid-tid. [Generic/Trade: solution 0.3% (5 mL). Trade only: ointment 0.3% (3.5 g tube).] ▶K ♀B ▶- $

Antibacterials - Fluoroquinolones

ciprofloxacin - ophthalmic (***Ciloxan***): 1-2 gtt q1-6h or ½ inch ribbon ointment bid-tid. [Generic/Trade: solution 0.3% (2.5,5,10 mL). Trade only: ointment 0.3% (3.5 g tube).] ▶LK ♀C ▶? $$

gatifloxacin - ophthalmic (***Zymar***): 1-2 gtts q2h while awake up to 8 times/day on days 1 & 2, then 1-2 gtts q4h up to 4 times/day on days 3-7. [Trade only: solution 0.3%.] ▶K ♀C ▶? $$$

levofloxacin - ophthalmic (***Iquix, Quixin***): Quixin: 1-2 gtts q2h while awake up to 8 times/day on days 1 & 2, then 1-2 gtts q4h up to 4 times/day on days 3-7. Iquix: 1-2 gtts q30 min to 2h while awake and q4-6h overnight on days 1-3, then 1-2 gtts q1-4h while awake on days 4 to completion of therapy. [Trade only: solution 0.5% (Quixin, 5 mL), 1.5% (Iquix, 5 mL).] ▶KL ♀C ▶? $$

moxifloxacin - ophthalmic (***Vigamox***): 1 gtt tid x 7 days. [Trade only: solution 0.5% (3 mL).] ▶LK ♀C ▶? $$$

ofloxacin - ophthalmic (***Ocuflox***): 1-2 gtts q1-6h x 7-10 days. [Generic/Trade: solution 0.3% (5 mL, 10 mL).] ▶LK ♀C ▶? $$

Antibacterials - Other

bacitracin - ophthalmic (*AK Tracin*): Apply ¼-½ inch ribbon of ointment q3-4h or bid-qid. [Generic/Trade: ointment 500 units/g 3.5g tube] ▶Min absorption ♀C ▶? $

erythromycin - ophthalmic (*Ilotycin, AK-Mycin*): ½ inch ribbon of ointment q3-4h or 2-8 times/day. [Generic: ointment 0.5% (1, 3.5 g tube).] ▶L ♀B ▶+ $

Neosporin ointment - ophthalmic (neomycin + bacitracin + polymyxin B): ½ inch ribbon of ointment q3-4h x 7-10 days or ½ inch ribbon 2-3 times/day for mild-moderate infection. [Generic/Trade: ointment. (3.5 g tube).] ▶K ♀C ▶? $$

Neosporin solution - ophthalmic (neomycin + polymyxin + gramicidin): 1-2 gtts q1-6h x 7-10 days. [Generic/Trade: solution (10 mL).] ▶KL ♀C ▶? $$

Polysporin - ophthalmic (polymyxin B + bacitracin): ½ inch ribbon of ointment q3-4h x 7-10 days or ½ inch ribbon bid-tid for mild-moderate infection. [Generic/Trade: ointment (3.5 g tube).] ▶K ♀C ▶? $$

Polytrim - ophthalmic (polymyxin B + trimethoprim): 1-2 gtts q3-6h x 7-10d, max 6 gtts/days. [Generic/Trade: Solution (10 mL).] ▶KL ♀C ▶? $

sulfacetamide - ophthalmic (*Sulamyd, Bleph-10, Sulf-10, Isopto Cetamide, AK-Sulf*): 1-2 gtts q2-6h x 7-10d or ½ inch ribbon of ointment q3-8h x 7-10d. [Generic only: solution 10%, 30% (15 mL). Generic/Trade: ointment 10% (3.5g tube). Trade only: solution 10% (2.5,5,15 mL) 10, 15, 30% (15 mL)] ▶K ♀C ▶- $

Antiviral Agents

trifluridine (*Viroptic*): Herpes: 1 gtt q2-4h x 7-14d, max 9 gtts/day, 21 days. [Trade only: solution 1% (7.5 mL).] ▶Minimal absorption ♀C ▶- $$$

vidarabine (*Vira-A*): ½ inch ribbon of ointment up to 5 times daily x 5-7 days. After re-epithelialization, ½ inch ribbon bid x 5-7 days. [Trade only: ointment 3% (3.5 g tube).] ▶Cornea ♀C ▶? $$$

Corticosteroid & Antibacterial Combinations

NOTE: Recommend that only ophthalmologists or optometrists prescribe due to infection, cataract, corneal/scleral perforation, and glaucoma risk. Monitor intraocular pressure.

Blephamide (prednisolone + sodium sulfacetamide): 1-2 gtts q1-8h or ½ inch ribbon of ointment daily-qid. [Generic/Trade: solution/suspension (5,10 mL), Trade only: ointment (3.5 g tube).] ▶KL ♀C ▶? $

Cortisporin - ophthalmic (neomycin + polymyxin + hydrocortisone): 1-2 gtts or ½ inch ribbon of ointment q3-4h or more frequently prn. [Generic/Trade: suspension (7.5 mL), ointment (3.5 g tube).] ▶LK ♀C ▶? $

FML-S Liquifilm (prednisolone + sodium sulfacetamide): 1-2 gtts q1-8h or ½ inch ribbon of ointment daily-qid. [Trade only: suspension (5,10 mL).] ▶KL ♀C ▶? $

Maxitrol (dexamethasone + neomycin + polymyxin): 1-2 gtts q1-8h or ½ -1 inch ribbon of ointment daily-qid. [Generic/Trade: suspension (5 mL), ointment (3.5 g tube).] ▶KL ♀C ▶? $

Pred G (prednisolone + gentamicin): 1-2 gtts q1-8h daily-qid or ½ inch ribbon of ointment bid-qid. [Trade only: suspension (2,5,10 mL), ointment (3.5 g tube).] ▶KL ♀C ▶? $$

TobraDex (tobramycin + dexamethasone): 1-2 gtts q2-6h or ½ inch ribbon of ointment bid-qid. [Trade: susp (2.5,5,10 mL), ointment (3.5 g tube).] ▶L ♀C ▶? $$$

Vasocidin (prednisolone + sodium sulfacetamide): 1-2 gtts q1-8h or ½ inch ribbon of ointment daily-qid. [Generic/Trade: solution (5,10 mL)] ▶KL ♀C ▶? $

Zylet (loteprednol + tobramycin): 1-2 gtts q1-2h x 1-2 days then 1-2 gtts q4-6h. [Trade: susp 0.5% loteprednol + 0.3% tobramycin (2.5,5,10 mL)] ▶LK ♀C ▶? $$$

Corticosteroids

NOTE: Recommend that only ophthalmologists or optometrists prescribe due to infection, cataract, corneal/scleral perforation, and glaucoma risk. Monitor intraocular pressure.

fluorometholone (***FML, FML Forte, Flarex, Fluor-Op***): 1-2 gtts q1-12h or ½ inch ribbon of ointment q4-24h. [Generic/Trade: suspension 0.1% (5,10,15 mL). Trade only: suspension 0.25% (5,10,15 mL), ointment 0.1% (3.5 g tube).] ▶L ♀C ▶? $$

loteprednol (***Alrex, Lotemax***): 1-2 gtts qid. [Trade only: suspension 0.2% (Alrex 5,10 mL), 0.5% (Lotemax 2.5, 5,10,15 mL).] ▶L ♀C ▶? $$

prednisolone - ophthalmic (***AK-Pred, Pred Forte, Pred Mild, Inflamase, Inflamase Forte, Econopred, Econopred Plus, ✦AK Tate, Diopred***): Solution: 1-2 gtts up to q1h during day and q2h at night, when response observed, then 1 gtt q4h, then 1 gtt tid-qid. Suspension: 1-2 gtts bid-qid. [Generic/Trade: suspension 1% (5,10, 15 mL), solution 0.125% (5 mL), 1% (5,10,15 mL). Trade only: suspension 0.12% (5,10 mL), solution 0.125% (5,10 mL).] ▶L ♀C ▶? $$

rimexolone (***Vexol***): 1-2 gtts q1-6h. [Trade only: susp 1% (5,10 mL.)] ▶L ♀C ▶? $$

Glaucoma Agents - Beta Blockers (Use caution in cardiac conditions and asthma.)

betaxolol - ophthalmic (***Betoptic, Betoptic S***): 1-2 gtts bid. [Trade only: suspension 0.25% (2.5,5,10,15 mL). Generic: solution 0.5% (5,10,15 mL).] ▶LK ♀C ▶? $$

carteolol - ophthalmic (***Ocupress***): 1 gtt bid. [Generic/Trade: solution 1% (5,10,15 mL).] ▶KL ♀C ▶? $

levobunolol (***Betagan***): 1- 2 gtts daily-bid. [Generic/Trade: solution 0.25% (5,10 mL) 0.5% (5,10,15 mL - Trade only 2 mL).] ▶? ♀C ▶- $

metipranolol (***Optipranolol***): 1 gtt bid. [Generic/Trade: solution 0.3% (5,10 mL).] ▶? ♀C ▶? $

timolol - ophthalmic (***Betimol, Timoptic, Timoptic XE, Istalol, Timoptic Ocudose***): 1 gtt bid. Timoptic XE, Istalol: 1 gtt daily. [Generic/Trade: solution 0.25, 0.5% (5,10,15 mL), preservative free 0.2 mL. Generic/Trade: gel forming solution 0.25, 0.5% (2.5*, 5 mL). Trade (Istalol): solution 0.5% 2.5, 5 mL. Note: *0.25% Timoptic XE.] ▶LK ♀C ▶+ $

Glaucoma Agents - Carbonic Anhydrase Inhibitors

NOTE: Sulfonamide derivatives; verify absence of sulfa allergy before prescribing.

acetazolamide (***Diamox, Diamox Sequels***): Glaucoma: 250 mg PO up to qid (immediate release) or 500 mg PO up to bid (extended release). Max 1 g/day. [Generic/Trade: Tabs 125,250 mg. Trade: ext'd release caps 500 mg.] ▶LK ♀C ▶+ $$

brinzolamide (***Azopt***): 1 gtt tid. [Trade only: susp 1% (5,10,15 mL.)] ▶LK ♀C ▶? $$

dorzolamide (***Trusopt***): 1 gtt tid. [Trade only: solution 2% (5,10 mL.)] ▶LK ♀C ▶- $$

methazolamide (***Neptazane***): 25-50 mg PO daily-tid. [Generic only: Tabs 25, 50 mg.] ▶LK ♀C ▶? $

Glaucoma Agents - Miotics

pilocarpine - ophthalmic (***Pilocar, Pilopine HS, Isopto Carpine, Ocusert, Ocu Carpine, ✦Diocarpine, Akarpine***): 1-2 gtts tid-qid up to 6 times/day or ½ inch ribbon of gel qhs. [Generic/Trade: solution 0.5, 1, 2, 3, 4, 6% (15mL). Trade only: 5%, 8% (15 mL), gel 4% (4 g tube).] ▶Plasma ♀C ▶? $

Glaucoma Agents - Prostaglandin Analogs

bimatoprost (***Lumigan***): 1 gtt qhs. [Trade only: solution 0.03% (2.5, 5, 7.5 mL).] ▶LK ♀C ▶? $$$

latanoprost (**Xalatan**): 1 gtt qhs. [Trade: soln 0.005% (2.5 mL).] ▶LK ♀C ▶? $$$
travoprost (**Travatan**): 1 gtt qhs. [Trade: soln 0.004% (2.5,5 mL).] ▶L ♀C ▶? $$$

Glaucoma Agents - Sympathomimetics

brimonidine (**Alphagan P, ✦Alphagan**): 1 gtt tid. [Trade only: solution 0.1% (5,10, 15 mL). Generic/Trade: solution 0.15% (5,10,15 mL). Generic only: 0.2% solution (5,10,15 mL).] ▶L ♀B ▶? $$

Glaucoma Agents - Other

Cosopt (dorzolamide + timolol): 1 gtt bid. [Trade only: solution dorzolamide 2% + timolol 0.5% (5, 7.5, 10, 18 mL).] ▶LK ♀D ▶- $$$

Mydriatics & Cycloplegics

atropine - ophthalmic (**Isopto Atropine**): 1-2 gtts before procedure or daily-qid, 1/8-1/4 inch ointment before procedure or daily-tid. Cycloplegia may last up to 5-10d and mydriasis may last up to 7-14d. [Generic/Trade: solution 1% (5,15 mL). Generic only: ointment 1% (3.5 g tube).] ▶L ♀C ▶+ $

cyclopentolate (**AK-Pentolate, Cyclogyl, Pentolair**): 1-2 gtts x 1-2 doses before procedure. Cycloplegia may last 6-24h; mydriasis may last 1 day. [Generic/Trade: soln 1% (2,15 mL). Trade only: 0.5% (15 mL) and 2% (2,5,15 mL).] ▶? ♀C ▶? $

homatropine (**Isopto Homatropine**): 1-2 gtts before procedure or bid-tid. Cycloplegia & mydriasis lasts 1-3 days. [Trade only: solution 2% (5 mL), 5% (15 mL). Generic/Trade: solution 5% (5 mL)] ▶? ♀C ▶? $

phenylephrine (**Neo-Synephrine, Mydfrin, Relief**): 1-2 gtts before procedure or tid-qid. No cycloplegia; mydriasis may last up to 5 hours. [Generic/Trade: solution 2.5,10% (3*, 5, 15 mL). Note: * 2.5% Mydfrin.] ▶Plasma, L ♀C ▶? $

tropicamide (**Mydriacyl**): 1-2 gtts before procedure. Mydriasis may last 6 hours. [Generic/Trade: solution 0.5% (15 mL), 1% (2, 3, 15 mL)] ▶? ♀? ▶? $

Nonsteroidal Anti-Inflammatories

bromfenac - ophthalmic (**Xibrom**): 1 gtt bid x 2 weeks. [Trade only: solution 0.09% (2.5, 5 mL).] ▶Minimal absorption ♀C, D (3rd trimester) ▶? $$$

diclofenac - ophthalmic (**Voltaren, ✦Voltaren Ophtha**): 1 gtt qid to qid. [Trade only: solution 0.1% (2.5,5 mL).] ▶L ♀B, D (3rd trimester) ▶? $$

ketorolac - ophthalmic (**Acular, Acular LS**): 1 gtt qid. [Trade only: solution Acular LS 0.4% (5 mL), Acular 0.5% (3,5,10 mL), preservative free Acular 0.5% unit dose (0.4 mL).] ▶L ♀C ▶? $$$

nepafenac (**Nevanac**): 1 gtt tid x 2 weeks. [Trade: suspension 0.1% (3 mL).] ▶Minimal absorption ♀C ▶? $$$

Other Ophthalmologic Agents

artificial tears (**Tears Naturale, Hypotears, Refresh Tears, GenTeal, Systane**): 1-2 gtts tid-qid prn. [OTC solution (15, 30 mL among others).] ▶Minimal absorption ♀A ▶+ $

cyclosporine - ophthalmic (**Restasis**): 1 gtt in each eye q12h. [Trade only: emulsion 0.05% (0.4 mL single-use vials).] ▶Minimal absorption ♀C ▶? $$$$

dapiprazole (**Rev-Eyes**): 2 gtt in each eye, repeat in 5 mins. [Trade only: powder for solution 0.5% (5 mL).] ▶Minimal absorption ♀B ▶? $

petrolatum (**Lacrilube, Dry Eyes, Refresh PM, ✦Duolube**): Apply ¼-½ inch ointment to inside of lower lid prn. [OTC ointment 3.5g tube.] ▶Minimal absorption ♀A ▶+ $

proparacaine (*Ophthaine, Ophthetic, ♥Alcaine*): 1-2 gtts before procedure. [Generic/Trade: solution 0.5% (15 mL).] ▶L ♀C ▶? $

tetracaine (*Pontocaine*): 1-2 gtts or ½-1 inch ribbon of ointment before procedure. [Generic/Trade: solution 0.5% (15 mL).] ▶Plasma ♀C ▶? $

> NOTE for proparacaine and tetracaine above: Do not prescribe for unsupervised or prolonged use. Corneal toxicity and ocular infections may occur with repeated use.

PSYCHIATRY

Antidepressants - Heterocyclic Compounds

amitriptyline (*Elavil*): Start 25-100 mg PO qhs; gradually increase to usual effective dose of 50-300 mg/d. Primarily inhibits serotonin reuptake. Demethylated to nortriptyline, which primarily inhibits norepinephrine reuptake. [Generic: Tabs 10, 25, 50, 75, 100,150 mg. Elavil brand name no longer available; has been retained in this entry for name recognition purposes only.] ▶L ♀D ▶- $$

clomipramine (*Anafranil*): Start 25 mg PO qhs; gradually increase to usual effective dose of 150-250 mg/d. Max 250 mg/d. Primarily inhibits serotonin reuptake. [Generic/Trade: Caps 25, 50, 75 mg.] ▶L ♀C ▶- $$

desipramine (*Norpramin*): Start 25-100 mg PO given once daily or in divided doses. Gradually increase to usual effective dose of 100-200 mg/d, max 300 mg/d. Primarily inhibits norepinephrine reuptake. [Generic/Trade: Tabs 10, 25, 50, 75, 100, 150 mg.] ▶L ♀C ▶+ $$

doxepin (*Sinequan*): Start 75 mg PO qhs. Gradually increase to usual effective dose of 75-150 mg/d, max 300 mg/d. Primarily inhibits norepinephrine. [Generic/Trade: Caps 10,25,50,75,100,150 mg. Oral concentrate 10 mg/mL.] ▶L ♀C ▶- $$

imipramine (*Tofranil, Tofranil PM*): Start 75-100 mg PO qhs or in divided doses; gradually increase to max 300 mg/d. Inhibits serotonin and norepinephrine reuptake. Demethylated to desipramine, which primarily inhibits norepinephrine reuptake. [Generic/Trade: Tabs 10, 25, 50 mg. Trade only: Caps 75, 100, 125, 150 mg (as pamoate salt).] ▶L ♀D ▶- $$$

nortriptyline (*Aventyl, Pamelor*): Start 25 mg PO given once daily or divided bid-qid. Usual effective dose is 75-100 mg/d, max 150 mg/d. Primarily inhibits norepinephrine reuptake. [Generic/Trade: Caps 10, 25, 50, 75 mg. Oral Solution 10 mg/5 mL.] ▶L ♀D ▶+ $$$

protriptyline (*Vivactil*): Depression: 15-40 mg/day PO divided tid-qid. Max 60 mg/day. [Trade only: Tabs 5, 10 mg.] ▶L ♀C ▶+ $$$

Antidepressants - Monoamine Oxidase Inhibitors (MAOIs)

> NOTE: May interfere with sleep; avoid qhs dosing. Must be on tyramine-free diet throughout treatment, and for 2 weeks after discontinuation. Numerous drug interactions; risk of hypertensive crisis and serotonin syndrome with many medications, including OTC. Allow ≥2 weeks wash-out when converting from an MAOI to an SSRI (6 weeks after fluoxetine), or other antidepressant.

isocarboxazid (*Marplan*): Start 10 mg PO bid; increase by 10 mg q2-4 days. Usual effective dose is 20-40 mg/d. MAOI diet. [Trade only: Tabs 10 mg.] ▶L ♀C ▶? $$$

phenelzine (*Nardil*): Start 15 mg PO tid. Usual effective dose is 60-90 mg/d in divided doses. MAOI diet. [Trade only: Tabs 15 mg.] ▶L ♀C ▶? $$$

selegiline (*Emsam*): Depression: start 6 mg/24hr patch q 24h. Max 12 mg/24h. MAOI diet for doses ≥ 9 mg/d. [Trade only: Transdermal patch 6 mg/24hr, 9 mg/24hr, 12 mg/24hr.] ▶L ♀C ▶? $$$$$

tranylcypromine (*Parnate*): Start 10 mg PO qam; increase by 10 mg/d at 1-3 wk intervals to usual effective dose of 10-40 mg/d divided bid. MAOI diet. [Generic/Trade: Tabs 10 mg.] ▶L ♀C ▶- $$

Antidepressants - Selective Serotonin Reuptake Inhibitors (SSRIs)

citalopram (*Celexa*): Depression: Start 20 mg PO daily; usual effective dose is 20-40 mg/d, max 60 mg/d. Suicidality. [Generic/Generic: Tabs 10, 20, 40 mg. Oral solution 10 mg/5 mL. Generic only: Oral disintegrating tab 10, 20, 40 mg.] ▶LK ♀C but - in 3rd trimester ▶- $$$

escitalopram (*Lexapro*): Depression, generalized anxiety disorder: Start 10 mg PO daily; max 20 mg/day. Suicidality. [Trade/Generic: Tabs 5, 10, 20 mg (trade 10 & 20 mg scored). Trade only: Oral solution 1 mg/mL.] ▶LK ♀C but - in 3rd trimester ▶- $$$

fluoxetine (*Prozac, Prozac Weekly, Sarafem*): Depression, OCD: Start 20 mg PO q am; usual effective dose is 20-40 mg/d, max 80 mg/d. Depression, maintenance: 20-40 mg/d (standard-release) or 90 mg PO once weekly (Prozac Weekly) starting 7 days after last standard-release dose. Bulimia: 60 mg PO daily; may need to titrate slowly, over several days. Panic disorder: Start 10 mg PO q am; titrate to 20 mg/d after one wk, max 60 mg/d. Premenstrual Dysphoric Disorder (Sarafem): 20 mg PO daily, given either throughout the menstrual cycle or for 14 days prior to menses; max 80 mg/d. Doses >20 mg/d can be divided bid (q am & q noon). Suicidality, many drug interactions. [Generic/Trade: Tabs 10 mg. Caps 10, 20, 40 mg. Oral solution 20 mg/5 mL. Trade only: Caps (Sarafem) 10, 15, 20 mg. Caps, delayed-release (Prozac Weekly) 90 mg. Generic only: Tabs 20, 40 mg.] ▶L ♀C but - in 3rd trimester ▶- $$$

fluvoxamine (♣*Luvox*): OCD: Start 50 mg PO qhs; usual effective dose is 100-300 mg/d divided bid, max 300 mg/d. OCD (children ≥8 yo): Start 25 mg PO qhs; usual effective dose is 50-200 mg/d divided bid, max 200 mg/d. Don't use with thioridazine, pimozide, alosetron, cisapride, tizanidine, tryptophan, or MAOIs; use caution with benzodiazepines, TCAs, theophylline, and warfarin. Suicidality. [Generic only: Tabs 25, 50, 100 mg.] ▶L ♀C but - in 3rd trimester ▶- $$$$

paroxetine (*Paxil, Paxil CR, Pexeva*): Depression: Start 20 mg PO qam, max 50 mg/d. Depression, controlled-release: Start 25 mg PO qam, max 62.5 mg/d. OCD: Start 10-20 mg PO qam, max 60 mg/d. Social anxiety disorder: Start 10-20 mg PO qam, max 60 mg/d. Social anxiety disorder, controlled-release: Start 12.5 mg PO qam, max 37.5 mg/d. Generalized anxiety disorder: Start 20 mg PO qam, max 50 mg/d. Panic disorder: Start 10 mg PO qam, increase by 10 mg/day at intervals ≥1 week to usual effective dose of 10-60 mg/d; max 60 mg/d. Panic disorder, controlled-release: Start 12.5 mg PO qam, max 75 mg/d. Post-traumatic stress disorder: Start 20 mg PO qam, max 50 mg/d. Premenstrual dysphoric disorder (PMDD), continuous dosing: Start 12.5 mg PO qam (controlled-release); may increase dose after 1 wk to max 25 mg qam. PMDD, intermittent dosing (given for 2 wks prior to menses): 12.5 mg PO qam (controlled-release), max 25 mg/d. Suicidality, many drug interactions. [Generic/Trade: Tabs 10, 20, 30, 40 mg. Trade only: Controlled-release tabs (Paxil CR) 12.5, 25, 37.5 mg. Oral suspension 10 mg/5 mL.] ▶LK ♀D ▶? $$$

sertraline (*Zoloft*): Depression, OCD: Start 50 mg PO daily; usual effective dose is 50-200 mg/d, max 200 mg/d. Panic disorder, post-traumatic stress disorder, social anxiety disorder: Start 25 mg PO daily, max 200 mg/d. Premenstrual dysphoric disorder (PMDD), continuous dosing: Start 50 mg PO daily, max 150 mg/d. PMDD, intermittent dosing (given for 14 days prior to menses): Start 50 mg PO daily x 3 days, then increase to 100 mg/d. Suicidality. [Generic/Trade: Tabs 25, 50, 100 mg. Oral concentrate 20 mg/mL.] ▶LK ♀C but - in 3rd trimester ▶+ $$$

Antidepressants - Serotonin-Norepinephrine Reuptake Inhibitors (SNRIs)

duloxetine (**Cymbalta**): Depression: 20 mg PO bid; max 60 mg/d given once daily or divided bid. Diabetic peripheral neuropathic pain: 60 mg PO daily. Suicidality, hepatotoxicity, many drug interactions. [Trade: Caps 20,30,60 mg] ▶L ♀C ▶? $$$$

venlafaxine (**Effexor, Effexor XR**): Depression/anxiety: Start 37.5-75 mg PO daily (Effexor XR) or 75 mg/d divided bid-tid (Effexor). Usual effective dose is 150-225 mg/d, max 225 mg/d (Effexor XR) or 375 mg/d (Effexor). Generalized anxiety disorder or social anxiety disorder: Start 37.5-75 mg PO daily (Effexor XR), max 225 mg/d. Panic disorder: Start 37.5 mg PO daily (Effexor XR), may titrate by 75 mg/d at weekly intervals to max 225 mg/d. Suicidality, seizures, hypertension. [Trade only: Caps, extended-release (Effexor XR) 37.5, 75, 150 mg. Tabs (Effexor) 25, 37.5, 50, 75, 100 mg.] ▶LK ♀C but - in 3rd trimester ▶? $$$$

Antidepressants - Other

bupropion (**Wellbutrin, Wellbutrin SR, Wellbutrin XL, Zyban, Buproban**): Depression: Start 100 mg PO bid (immediate-release tabs); can increase to 100 mg tid after 4-7 d. Usual effective dose is 300-450 mg/d, max 150 mg/dose and 450 mg/d. Sustained-release: Start 150 mg PO q am; may increase to 150 mg bid after 4-7 d, max 400 mg/d. Give last dose no later than 5 pm. Extended-release: Start 150 mg PO q am; may increase to 300 mg q am after 4 d, max 450 mg q am. Seasonal affective disorder: Start 150 mg of extended release PO q am in autumn; can increase to 300 mg q am after 1 wk, max 300 mg/d. In the spring, decrease to 150 mg/d for 2 weeks and then discontinue. Smoking cessation (Zyban, Buproban): Start 150 mg PO q am x 3d, then increase to 150 mg PO bid x 7-12 wks. Max 150 mg PO bid. Give last dose no later than 5 pm. Seizures, suicidality. [Generic/Trade (for depression): Tabs 75,100 mg. Sustained release tabs 100, 150, 200 mg. Trade only: Extended-release tabs 150, 300 mg (Wellbutrin XL). Generic/Trade (smoking cessation): Sustained-release tabs 150 mg (Zyban, Buproban).] ▶LK ♀C ▶- $$$$

mirtazapine (**Remeron, Remeron SolTab**): Start 15 mg PO qhs. Usual effective dose is 15-45 mg/d. Agranulocytosis in 0.1% of patients. Suicidality. [Generic/Trade: Tabs 15, 30, 45 mg. Tabs, orally disintegrating (SolTab) 15, 30, 45 mg. Generic: Tabs 7.5 mg.] ▶LK ♀C ▶? $$

nefazodone: Start 100 mg PO bid (50 mg/d in elderly or debilitated patients). Usual effective dose is 150-300 mg PO bid, max 600 mg/d. Many drug interactions; don't use with cisapride, MAOIs, pimozide, or triazolam. Hepatotoxicity, suicidality. [Generic only: Tabs 50, 100, 150, 200, 250 mg.] ▶L ♀C ▶? $$$

trazodone (**Desyrel**): Depression: Start 50-150 mg PO in divided doses; usual effective dose is 400-600 mg/d. Insomnia: 50-150 mg PO qhs. [Generic/Trade: Tabs 50, 100, 150, 300 mg.] ▶L ♀C ▶- $$$$

Antimanic (Bipolar) Agents

carbamazepine (**Tegretol, Tegretol XR, Carbatrol, Epitol, Equetro**): Bipolar disorder, acute manic/mixed episodes (Equetro): Start 200 mg PO bid, increase by 200 mg/d to max 1,600 mg/d. Aplastic anemia, many drug interactions. [Generic/Trade: Tabs 200 mg, chew tabs 100 mg, susp 100 mg/5 mL. Trade only: Extended release tabs (Tegretol XR): 100, 200, 400 mg. Extended release caps (Equetro, Carbatrol): 100, 200, 300 mg. Generic Only: Chew tabs 200 mg, tabs 100, 300, 400 mg.] ▶LK ♀D ▶+ $$

lamotrigine (**Lamictal, Lamictal CD**): Adults with bipolar disorder (maintenance):

Start 25 mg PO daily, 50 mg PO daily if on enzyme-inducing drugs, or 25 mg PO qod if on valproate; titrate to 200 mg/d, 400 mg/d divided bid if on enzyme-inducing drugs, or 100 mg/d if on valproate. Potentially life-threatening rashes in 0.3% of adults and 0.8% of children; discontinue at first sign of rash. Drug interaction with valproic acid; see product information for adjusted dosing guidelines. [Generic/Trade: Chewable dispersible tabs 5, 25 mg. Trade only: Tabs 25, 100, 150, 200 mg.] ▶LK ♀C ▶- $$$$

lithium (*Eskalith, Eskalith CR, Lithobid,* ♣*Lithane*): Acute mania: Start 300-600 mg PO bid-tid. Usual effective dose is 900-1800 mg/d. Steady state is achieved in 5d. Usual therapeutic trough levels are 1.0-1.5 mEq/L (acute mania) or 0.6-1.2 mEq/L (maintenance). 300 mg = 8 mEq or mmol. [Generic/Trade: Caps 300, Extended release tabs 300, 450 mg. Generic only: Caps 150, 600 mg, Tabs 300 mg, Syrup 300/5 mL.] ▶K ♀D ▶- $

topiramate (*Topamax*): Bipolar disorder (unapproved): Start 25-50 mg/d PO. Titrate prn to max 400 mg/d divided bid. [Trade only: Tabs 25, 50, 100, 200 mg. Sprinkle caps 15, 25 mg.] ▶K ♀C ▶? $$$$$

valproic acid (*Depakote, Depakote ER, divalproex,* ♣*Epiject, Epival, Deproic*): Mania: 250 mg PO tid (Depakote) or 25 mg/kg once daily (Depakote ER); max 60 mg/kg/d. Hepatotoxicity, drug interactions, reduce dose in the elderly. [Generic/Trade: Caps 250mg (Depakene), syrup (Depakene, valproic acid) 250 mg/5 mL. Trade only (Depakote): Caps, sprinkle 125 mg, delayed release tabs 125, 250, 500 mg; extended release tabs (Depakote ER) 250, 500mg.] ▶L ♀D ▶+ $$$$

Antipsychotics - Atypical - Serotonin Dopamine Receptor Antagonists

clozapine (*Clozaril, FazaClo ODT*): Start 12.5 mg PO daily or bid. Usual effective dose is 300-450 mg/d divided bid, max 900 mg/d. Agranulocytosis 1-2%; check WBC and ANC q week x 6 m, then q2 wks. Seizures, myocarditis, cardiac arrest. [Generic/Trade: Tabs 25, 100 mg. Generic only: Tabs 12.5, 50, 200 mg. Trade: Orally disintegrating tab (Fazaclo ODT) 25, 100 mg (scored).] ▶L ♀B ▶- $$$$$

olanzapine (*Zyprexa, Zyprexa Zydis*): Agitation in acute bipolar mania or schizophrenia: Start 10 mg IM (2.5-5 mg in elderly or debilitated patients); may repeat in ≥2h to max 30 mg/d. Psychotic disorders, oral therapy: Start 5-10 mg PO daily; usual effective dose is 10-15 mg/d. Bipolar disorder, maintenance treatment or monotherapy for acute manic or mixed episodes: Start 10-15 mg PO daily. Increase by 5 mg/d at intervals ≥24 h to usual effective dose of 5-20 mg/d, max 20 mg/d. Bipolar disorder, adjunctive for acute manic or mixed episodes: Start 10 mg PO daily; usual effective dose is 5-20 mg/d, max 20 mg/d. [Trade only: Tabs 2.5, 5, 7.5, 10, 15, 20 mg. Tabs, orally-disintegrating (Zyprexa Zydis) 5,10,15, 20 mg.] ▶L ♀C ▶- $$$$$

quetiapine (*Seroquel*): Schizophrenia: Start 25 mg PO daily; increase by 25-50 mg bid-tid on days 2 and 3, and then to target dose of 300-400 mg/d divided bid-tid on day 4. Usual effective dose is 150-750 mg/d, max 800 mg/d. Acute bipolar mania: Start 50 mg PO bid on day 1, then increase to no higher than 100 mg bid on day 2, 150 mg bid on day 3, and 200 mg bid on day 4. May increase prn to 300 mg bid on day 5 and 400 mg bid thereafter. Usual effective dose is 400-800 mg/d. Eye exam for cataracts recommended q 6 m. [Trade only: Tabs 25, 50, 100, 200, 300, 400 mg.] ▶LK ♀C ▶- $$$$$

risperidone (*Risperdal, Risperdal Consta*): Psychotic disorders: Start 1 mg PO bid (0.5 mg in the elderly); slowly increase to usual effective dose of 4-8 mg/d given once daily or divided bid, max 16 mg/d. Long-acting injection (Consta): Start

25 mg IM q 2 wks while continuing oral dose x 3 wks. May increase at 4 wk intervals to max 50 mg q 2 wks. Bipolar mania: Start 2-3 mg PO daily; may increase by 1 mg/day at 24 hr intervals to max 6 mg/d. [Trade only: Tabs 0.25, 0.5, 1, 2, 3, 4 mg. Orally disintegrating tablets (M-TAB) 0.5, 1, 2 mg. Oral solution 1 mg/mL.] ▶LK ♀C ▶- $$$$$

ziprasidone (Geodon): Schizophrenia: Start 20 mg PO bid with food; may adjust at >2 d intervals to max 80 mg PO bid. Acute agitation: 10-20 mg IM, max 40 mg/d. Bipolar mania: Start 40 mg PO bid with food; may increase to 60-80 mg bid on day 2. Usual effective dose is 40-80 mg bid. [Trade only: Caps 20, 40, 60, 80 mg, Susp 10 mg/mL.] ▶L ♀C ▶- $$$$$

Antipsychotics - D2 Antagonists - High Potency (1-5 mg = 100 mg CPZ)

fluphenazine (Prolixin, ♥Modecate, Modeten): 1.25-10 mg/d IM divided q6-8h. Start 0.5-10 mg/d PO divided q6-8h. Usual effective dose 1-20 mg/d PO. Depot (fluphenazine decanoate/ enanthate): 12.5-25 mg IM/SC q3 weeks = 10-20 mg/d PO fluphenazine. [Generic/Trade: Tabs 1, 2.5, 5, 10 mg. Elixir 2.5 mg/5 mL. Oral concentrate 5 mg/mL.] ▶LK ♀C ▶? $$$

haloperidol (Haldol): 2-5 mg IM. Start 0.5- 5 mg PO bid-tid, usual effective dose 6-20 mg/d. Therapeutic range 2-15 ng/mL. Depot haloperidol (haloperidol decanoate): 100-200 mg IM q4 weeks = 10 mg/day oral haloperidol. [Generic only: Tabs 0.5, 1, 2, 5, 10, 20 mg. Oral concentrate 2 mg/mL.] ▶LK ♀C ▶- $$$

perphenazine (Trilafon): Start 4-8 mg PO tid or 8-16 mg PO bid-qid (hospitalized patients), maximum 64 mg/d. Can give 5-10 mg IM q6h, max 30 mg/day IM. [Generic: Tabs 2, 4, 8, 16 mg. Oral concentrate 16 mg/5 mL.] ▶LK ♀C ▶? $$$

pimozide (Orap): Tourette's: Start 1-2 mg PO in divided doses, increase q2 days to usual effective dose of 1-10 mg/d. [Trade only: Tabs 1, 2 mg.] ▶L ♀C ▶? $$$

thiothixene (Navane): Start 2 mg PO tid. Usual effective dose is 20-30 mg/d, maximum 60 mg PO. [Generic/Trade: Caps 1, 2, 5, 10. Oral concentrate 5 mg/mL. Trade only: Caps 20 mg.] ▶LK ♀C ▶? $$$

trifluoperazine (Stelazine): Start 2-5 mg PO bid. Usual effective dose is 15- 20 mg/d. Can give 1-2 mg IM q4-6h prn, maximum 10 mg/d IM. [Generic/Trade: Tabs 1, 2, 5, 10 mg. Trade only: Oral concentrate 10 mg/mL.] ▶LK ♀C ▶- $$$

Antipsychotics - D2 Antagonists - Low Potency (50-100 mg = 100 mg CPZ)

chlorpromazine (Thorazine, ♥Largactil): Start 10-50 mg PO/IM bid-tid, usual dose 300-800 mg/d. [Generic only: Tabs 10, 25, 50, 100, 200 mg. Generic/Trade: Oral concentrate 30 mg/mL, 100 mg/mL. Trade only: Syrup 10 mg/5 mL. Suppositories 25, 100 mg.] ▶LK ♀C ▶- $$$

thioridazine (Mellaril, ♥Rideril): Start 50-100 mg PO tid, usual dose 200-800 mg/d. Not first-line therapy. Causes QTc prolongation, torsade de pointes, sudden death. Contraindicated with SSRIs, propranolol, pindolol. Monitor baseline ECG and potassium. Pigmentary retinopathy with doses>800 mg/d. [Generic: Tabs 10,15,25,50,100,150,200 mg. Oral concentrate 30,100 mg/mL.] ▶LK ♀C ▶? $$

Antipsychotics - Dopamine-2/serotonin-1 partial agonist & ser-2 antagonist

aripiprazole (Abilify, Abilify Discmelt): Schizophrenia: Start 10-15 mg PO daily. Max 30 mg daily. Bipolar disorder: Start 30 mg PO daily; reduce dose to 15 mg/day if higher dose poorly tolerated. [Trade only: Tabs 2,5,10,15,20,30 mg. Oral solution 1 mg/mL (150 mL). Orally disintegrating tabs (Discmelt) 10, 15, 20, 30 mg.] ▶L ♀C ▶? $$$$$

Anxiolytics / Hypnotics - Benzodiazepines - Long Half-Life (25-100 hours)

bromazepam (❦*Lectopam*): Canada only. 6-18 mg/d PO in divided doses. [Generic/Trade: Tabs 1.5, 3, 6 mg.] ▶L ♀D ▶- $

chlordiazepoxide (*Librium*): Anxiety: 5-25 mg PO or 25-50 mg IM/IV tid-qid. Acute alcohol withdrawal: 50-100 mg PO/IM/IV, repeat q3-4h prn up to 300 mg/d. Half-life 5-30 h. [Generic/Trade: Caps 5, 10, 25 mg.] ▶LK ♀D ▶- ©IV $$

clonazepam (*Klonopin, Klonopin Wafer,* ❦*Rivotril, Clonapam*): Start 0.25-0.5 mg PO bid-tid, max 4 mg/day. Half-life 18- 50 h. [Generic/Trade: Tabs 0.5, 1, 2 mg. Orally disintegrating tabs 0.125, 0.25, 0.5, 1, 2 mg.] ▶LK ♀D ▶? ©IV $$

clorazepate (*Tranxene, Tranxene SD, Genxene*): Start 7.5-15 mg PO bid- or bid-tid, usual effective dose is 15-60 mg/day. Acute alcohol withdrawal: 60-90 mg/day on first day divided bid-tid, reduce dose to 7.5-15 mg/day over 5 days. [Generic/Trade: Tabs 3.75, 7.5, 15 mg. Trade only (Tranxene SD): Extended release Tabs 11.25, 22.5 mg.] ▶LK ♀D ▶- ©IV $$$$

diazepam (*Valium,* ❦*Vivol, E Pam, Diazemuls*): Anxiety: 2-10 mg PO bid-qid. Half-life 20-80h. Alcohol withdrawal: 10 mg PO tid-qid x 24 hr then 5 mg PO tid-qid prn. [Generic/Trade: Tabs 2, 5, 10 mg. Generic only: Oral solution 5 mg/5 mL. Oral concentrate (Intensol) 5 mg/mL.] ▶LK ♀D ▶- ©IV $

flurazepam (*Dalmane*): 15-30 mg PO qhs. Half-life 70-90h. [Generic/Trade: Caps 15, 30 mg.] ▶LK ♀X ▶- ©IV $

Anxiolytics / Hypnotics - Benzodiazepines - Medium Half-Life (10-15 hours)

estazolam (*ProSom*): 1-2 mg PO qhs. [Generic/Trade: Tabs 1, 2 mg.] ▶LK ♀X ▶- ©IV $$

lorazepam (*Ativan*): 0.5-2 mg IV/IM/PO q6-8h, max 10 mg/d. Half-life 10-20h. [Generic only: Oral concentrate 2 mg/mL.] ▶LK ♀D ▶- ©IV $$$

temazepam (*Restoril*): 7.5-30 mg PO qhs. Half-life 8-25h. [Generic/Trade: Caps 15, 30 mg. Trade only: Caps 7.5, 22.5 mg.] ▶LK ♀X ▶- ©IV $

Anxiolytics / Hypnotics - Benzodiazepines - Short Half-Life (<12 hours)

alprazolam (*Xanax, Xanax XR, Niravam*): 0.25-0.5 mg PO bid-tid. Half-life 12h. Multiple drug interactions. [Trade only: Orally disintegrating tab (Niravam) 0.25, 0.5, 1, 2 mg. Generic/Trade: Tabs 0.25, 0.5, 1, 2 mg. Extended release tabs: 0.5, 1, 2, 3 mg. Generic only: Oral concentrate (Intensol) 1 mg/mL.] ▶LK ♀D ▶- ©IV $

oxazepam (*Serax*): 10-30 mg PO tid-qid. Half-life 8h. [Generic/Trade: Caps 10, 15, 30 mg. Trade only: Tabs 15 mg.] ▶LK ♀D ▶- ©IV $$$

triazolam (*Halcion*): 0.125-0.5 mg PO qhs. 0.125 mg/day in elderly. Half-life 2-3h. [Generic/Trade: Tabs 0.125, 0.25 mg.] ▶LK ♀X ▶- ©IV $

Anxiolytics / Hypnotics - Other

buspirone (*BuSpar, Vanspar*): Anxiety: Start 15 mg "dividose" daily (7.5 mg PO bid), usual effective dose 30 mg/day. Max 60 mg/day. [Generic/Trade: Tabs 5, 10, 15 mg. Trade only: Dividose tab 15, 30 mg (scored to be easily bisected or trisected). Generic only: Tabs 7.5 mg.] ▶K ♀B ▶- $$$

chloral hydrate (*Aquachloral Supprettes, Somnote*): 25-50 mg/kg/day up to 1000 mg PO/PR. Many physicians use higher than recommended doses in children (eg, 75 mg/kg). [Generic only: Syrup 500 mg/5 mL, rectal suppositories 500 mg. Trade only: Caps 500 mg. Rectal suppositories: 325, 650 mg.] ▶LK ♀C ▶+ ©IV $

eszopiclone (*Lunesta*): 2 mg PO qhs prn. Max 3 mg. Elderly: 1 mg PO qhs prn, max 2 mg. [Trade only: Tabs 1, 2, 3 mg.] ▶L ♀C ▶? $$$$

ramelteon (*Rozerem*): Insomnia: 8 mg PO qhs. [Trade: tabs 8 mg.] ▶L ♀C ▶? $$$

zaleplon (*Sonata*, ✚*Starnoc*): 5-10 mg PO qhs prn, max 20 mg. Do not use for benzodiazepine or alcohol withdrawal. [Trade: Caps 5, 10 mg.] ▶L ♀C ▶- ©IV $$$

zolpidem (*Ambien, Ambien CR*): 5-10 mg PO qhs (standard tabs) or 6.25-12.5 mg PO qhs (controlled release tabs). Do not use for benzodiazepine or alcohol withdrawal. [Trade only: Tabs 5,10 mg. Controlled release tabs 6.25, 12.5 mg.] ▶L ♀B ▶+ ©IV $$$$

zopiclone (✚*Imovane*): Canada only. Adults: 5-7.5 mg PO qhs. Reduce dose in elderly. [Trade only: tabs 5, 7.5 mg. Generic: tabs 7.5 mg.] ▶L ♀D ▶- $

Combination Drugs

Symbyax (olanzapine + fluoxetine): Bipolar depression: Start 6/25 mg PO qhs. Max 18/75 mg/day. [Trade only: Caps (olanzapine/fluoxetine) 6/25, 6/50, 12/25, 12/50 mg.] ▶LK ♀C ▶- $$$$$

Drug Dependence Therapy

acamprosate (*Campral*): Maintenance of abstinence from alcohol: 666 mg (2 tabs) PO tid. Start after alcohol withdrawal and when patient is abstinent. [Trade only: delayed-release tabs 333 mg.] ▶K ♀C ▶? $$$$

buprenorphine (*Subutex*): Treatment of opioid dependence: Induction 8 mg SL on day 1, 16 mg SL on day 2. Maintenance: 16 mg SL daily. Can individualize to range of 4-24 mg SL daily. [Trade only: SL tabs 2, 8 mg.] ▶L ♀C ▶- ©III $$$$

disulfiram (*Antabuse*): Sobriety: 125-500 mg PO daily. Patient must abstain from any alcohol for ≥12 h before using. Metronidazole and alcohol in any form (cough syrups, tonics, etc.) contraindicated. [Trade only: Tabs 250 mg.] ▶L ♀C ▶? $$

methadone (*Dolophine, Methadose*, ✚*Metadol*): Opioid dependence: 20-100 mg PO daily. Treatment >3 wks is maintenance and only permitted in approved treatment programs. [Generic/Trade: Tabs 5, 10, 40 mg. Oral solution 5 & 10 mg/5 mL. Oral concentrate 10 mg/mL.] ▶L ♀B ▶+ ©II $$

naltrexone (*ReVia, Depade, Vivitrol*): Alcohol/opioid dependence: 25-50 mg PO daily. Avoid if recent ingestion of opioids (past 7-10 days). Hepatotoxicity with higher than approved doses. [Generic/Trade: Tabs 50 mg. Trade only (Vivitrol): extended-release injectable suspension kits 380 mg.] ▶LK ♀C ▶? $$$$

nicotine gum (*Nicorette, Nicorette DS*): Smoking cessation: Gradually taper 1 piece q1-2h x 6 weeks, 1 piece q2-4h x 3 weeks, then 1 piece q4-8h x 3 weeks, max 30 pieces/day of 2 mg or 24 pieces/day of 4 mg. Use Nicorette DS 4 mg/ piece in high cigarette use (>24 cigarettes/day). [OTC/Generic/Trade: gum 2, 4 mg.] ▶LK ♀X ▶- $$$$$

nicotine inhalation system (*Nicotrol Inhaler*, ✚*Nicorette inhaler*): 6-16 cartridges/ day x 12 weeks [Trade only: Oral inhaler 10 mg/cartridge (4 mg nicotine delivered), 42 cartridges/box.] ▶LK ♀D ▶- $$$$$

nicotine lozenge (*Commit*): Smoking cessation: In those who smoke <30 min from waking use 4 mg lozenge; others use 2 mg. Take 1-2 lozenges q1-2 h x 6 weeks, then q2-4h in weeks 7-9, then q4-8h in weeks 10-12. Length of therapy 12 weeks. [OTC Generic/Trade: lozenge 2,4 mg in 72,168-count packages] ▶LK ♀D ▶- $$$$$

nicotine nasal spray (*Nicotrol NS*): Smoking cessation:1-2 doses each hour, with each dose = 2 sprays, one in each nostril (1 spray = 0.5 mg nicotine). Minimum recommended: 8 doses/day, max 40 doses/day. [Trade only: nasal solution 10 mg/mL (0.5 mg/inhalation); 10 mL bottles.] ▶LK ♀D ▶- $$$$$

nicotine patches (**Habitrol, Nicoderm, Nicotrol, ✦Prostep**): Smoking cessation: Start one patch (14-22 mg) daily, taper after 6 wks. Ensure patient has stopped smoking. [OTC/Rx/Generic/Trade: patches 11, 22 mg/ 24 hours. 7, 14, 21 mg/ 24 h (Habitrol & NicoDerm). OTC/Trade: 15 mg/ 16 h (Nicotrol).] ▶LK ♀D ▶- $$$$

Suboxone (buprenorphine + naloxone): Treatment of opioid dependence: Maintenance: 16 mg SL daily. Can individualize to range of 4-24 mg SL daily. [Trade only: SL tabs 2/0.5 and 8/2 mg buprenorphine / naloxone] ▶L ♀C ▶- ©III $$$$$

varenicline (**Chantix**): Smoking cessation: Start 0.5 mg PO daily for days 1-3, then 0.5 mg bid days 4-7, then 1 mg bid thereafter. Take after meals with full glass of water. Start 1 wk prior to cessation and continue x 12 wks. [Trade: Tabs 0.5, 1 mg.] ▶K ♀C ▶? ?

Stimulants / ADHD / Anorexiants

Adderall (dextroamphetamine + amphetamine, Adderall XR): ADHD, standard-release tabs: Start 2.5 mg (3-5 yo) or 5 mg (≥6 yo) PO daily-bid, increase by 2.5-5 mg every week, max 40 mg/d. ADHD, extended-release caps (Adderall XR): If 6-12 yo, then start 5-10 mg PO daily to a max of 30 mg/d. If 13-17 yo, then start 10 mg PO daily to a max of 20 mg/d. If adult, then 20 mg PO daily. Narcolepsy, standard-release: Start 5-10 mg PO q am, increase by 5-10 mg q week, max 60 mg/d. Avoid evening doses. Monitor growth and use drug holidays when appropriate. [Generic/Trade: Tabs 5, 7.5, 10, 12.5, 15, 20, 30 mg. Trade only: Capsules, extended release (Adderall XR) 5, 10, 15, 20, 25, 30 mg.] ▶L ♀C ▶- ©III $$$$

atomoxetine (**Strattera**): ADHD: Children/adolescents >70 kg and adults: Start 40 mg PO daily, then increase after >3 days to target of 80 mg/d divided daily-bid. Max 100 mg/d. [Trade: caps 10, 18, 25, 40, 60, 80, 100 mg.] ▶K ♀C ▶? $$$$$

caffeine (**NoDoz, Vivarin, Caffedrine, Stay Awake, Quick-Pep**): 100-200 mg PO q3-4h prn. [OTC/Generic/Trade: Tabs/Caps 200 mg. OTC/Trade: Extended-release tabs 200 mg. Lozenges 75 mg.] ▶L ♀B ▶? $

dexmethylphenidate (**Focalin, Focalin XR**): ADHD, extended release, not already on stimulants: 5 mg (children) or 10 mg (adults) PO q am. Immediate release, not already on stimulants: 2.5 mg PO bid. Max 20 mg/day for both. If taking racemic methylphenidate use conversion of 2.5 mg for each 5 mg of methylphenidate, max 20 mg/d. [Trade only: Immediate release tabs 2.5, 5, 10 mg. Extended release caps (Focalin XR) 5, 10, 20 mg.] ▶LK ♀C ▶? ©II $$

dextroamphetamine (**Dexedrine, Dextrostat**): Narcolepsy/ ADHD: 2.5-10 mg PO q am or bid-tid or 10-15 mg PO daily (sustained release), max 60 mg/d. Avoid evening doses. Monitor growth and use drug holidays when appropriate. [Generic/Trade: Tabs 5, 10 mg. Extended Release caps 5, 10, 15 mg.] ▶L ♀C ▶- ©II $$$

methylphenidate (**Ritalin, Ritalin LA, Ritalin SR, Methylin, Methylin ER, Metadate ER, Metadate CD, Concerta, Daytrana**): ADHD/Narcolepsy: 5-10 mg PO bid-tid or 20 mg PO qam (sustained and extended release), max 60 mg/d. Or 18-36 mg PO qam (Concerta), max 72 mg/day. Avoid evening doses. Monitor growth and use drug holidays when appropriate. [Trade only: tabs 5, 10, 20 mg (Ritalin, Methylin, Metadate). Extended release tabs 10, 20 mg (Methylin ER, Metadate ER). Extended release tabs 18, 27, 36, 54 mg (Concerta). Extended release caps 10, 20, 30, 40, 50, 60 mg (Metadate CD) May be sprinkled on food. Sustained-release tabs 20 mg (Ritalin SR). Extended release caps 10, 20, 30, 40 mg (Ritalin LA). Chewable tabs 2.5, 5, 10 mg (Methylin). Oral soln 5mg/5ml, 10 mg/5 ml (Methylin). Transdermal patch (Daytrana) 10 mg/9 hrs, 15 mg/9 hrs, 20 mg/9 hrs, 30 mg/9 hrs. Generic: tabs 5, 10, 20 mg, extended release tabs 10, 20 mg, sustained-release tabs 20 mg.] ▶LK ♀C ▶? ©II $$

modafinil (**Provigil**, ♦**Alertec**): Narcolepsy and sleep apnea/hypopnea: 200 mg PO qam. Shift work sleep disorder: 200 mg PO one hour before shift. [Trade only: Tabs 100, 200 mg.] ▶L ♀C ▶- ©IV $$$$

phentermine (**Adipex-P, Ionamin, Pro-Fast**): 8 mg PO tid or 15-37.5 mg/day q am or 10-14 h before retiring. For short-term use. [Generic/Trade: Caps 15, 18.75, 30, 37.5 mg. Tabs 8, 30, 37.5 mg. Trade only: extended release Caps 15, 30 mg (Ionamin).] ▶KL ♀C ▶- ©IV $$

sibutramine (**Meridia**): Start 10 mg PO q am, max 15 mg/d. Monitor pulse and BP. [Trade only: Caps 5, 10, 15 mg.] ▶KL ♀C ▶- ©IV $$$$

BODY MASS INDEX*	Heights are in feet and inches; weights are in pounds					
BMI	_Classification_	_4'10"_	_5'0"_	_5'4"_	_5'8"_	_6'0"_	_6'4"_
<19	Underweight	<91	<97	<110	<125	<140	<156
19-24	Healthy Weight	91-119	97-127	110-144	125-163	140-183	156-204
25-29	Overweight	120-143	128-152	145-173	164-196	184-220	205-245
30-40	Obese	144-191	153-204	174-233	197-262	221-293	246-328
>40	Very Obese	>191	>204	>233	>262	>293	>328

*BMI = kg/m^2 = (weight in pounds)(703)/(height in inches)^2. Anorectants appropriate if BMI ≥30 with comorbidities ≥27); surgery an option if BMI >40 with comorbidities 35-40).

www.nhlbi.nih.gov

Other Agents

clonidine (**Catapres, Catapres-TTS**): ADHD (unapproved peds): Start 0.05 mg qhs, titrate based on response over 8 weeks to max 0.2 mg/day (<45 kg) or 0.4 mg/day (> 45 kg) in 2-4 divided doses. Tourette's syndrome (unapproved peds/adult): 3-5 mcg/kg/d PO divided bid-qid. Opioid/nicotine/alcohol withdrawal adjunct (not FDA approved): 0.1- 0.3 mg PO tid-qid. Transdermal Therapeutic System (TTS) is designed for 7 day use so that a TTS-1 delivers 0.1 mg/day x 7 days. May supplement first dose of TTS with oral x 2-3 days while therapeutic level is achieved. [Generic/Trade: Tabs 0.1, 0.2, 0.3 mg. Trade only: transdermal weekly patch 0.1 mg/day (TTS-1), 0.2 mg/day (TTS-2), 0.3 mg/day (TTS-3).] ▶LK ♀C ▶? $

diphenhydramine (**Benadryl, Banophen, Allermax, Diphen, Diphenhist, Siladryl, Sominex**, ♦**Allerdryl, Allernix**): EPS: 25-50 mg PO tid-qid or 10-50 mg IV/IM tid-qid. Insomnia: 25-50 mg PO qhs. Peds ≥12 yo: 50 mg PO qhs. [OTC/Generic/Trade: Tabs 25, 50 mg. OTC/Trade only: chew Tabs 12.5 mg. OTC/Rx/Generic/Trade: Caps 25, 50 mg. oral solution 12.5 mg/5 mL.] ▶LK ♀B ▶- $

PULMONARY

Beta Agonists

albuterol (**Ventolin, Ventolin HFA, Proventil, Proventil HFA, ProAir HFA, Volmax, VoSpire ER, Ventodisk**, ♦**Airomir, Ventodisk, Asmavent, salbutamol**): MDI 2 puffs q4-6h prn. 0.5 mL of 0.5% soln (2.5 mg) nebulized tid-qid. One 3 mL unit dose (0.083%) nebulized tid-qid. Caps for inhalation 200-400 mcg q4-6h. 2-4 mg PO tid-qid or extended release 4-8 mg PO q12h up to 16 mg PO q12h. Children 2-5 yo: 0.1-0.2 mg/kg/dose PO tid up to 4 mg tid; 6-12 yo: 2-4 mg or extended release 4 mg PO q12h. Prevention of exercise-induced bronchospasm: MDI: 2 puffs 15-30 minutes before exercise. [Generic/Trade: MDI 90 mcg/actuation, 17g-200/canister. "HFA" inhalers use hydrofluoroalkane propellant instead of CFCs but are otherwise equivalent. Soln for inhalation 0.042% and 0.083% in 3 mL vial, 0.5% (5 mg/mL) in 20 mL with dropper. Tabs 2, 4 mg. Generic only: Syrup 2 mg/5 mL. Trade only: (AccuNeb) Soln for inhalation 0.021% in 3 mL vial. Extended release tabs 4, 8 mg.] ▶L ♀C ▶? $

PREDICTED PEAK EXPIRATORY FLOW (liters/min)   *Am Rev Resp Dis* 1963; 88:644

Age (yrs)	Women (height in inches)					Men (height in inches)					Child (height in inches)	
	55"	60"	65"	70"	75"	60"	65"	70"	75"	80"		
20	390	423	460	496	529	554	602	649	693	740	44"	160
30	380	413	448	483	516	532	577	622	664	710	46"	187
40	370	402	436	470	502	509	552	596	636	680	48"	214
50	360	391	424	457	488	486	527	569	607	649	50"	240
60	350	380	412	445	475	463	500	542	578	618	52"	267
70	340	369	400	432	461	440	477	515	550	587	54"	293

fenoterol (♣*Berotec*): Canada only. 1-2 puffs prn tid-qid. Nebulizer: up to 2.5 mg q6 hours. [Trade only: MDI 100 mcg/actuation. Soln for inhalation: 20 mL bottles of 1 mg/mL (with preservatives that may cause bronchoconstriction in those with hyperreactive airways).] ▶L ♀C ▶? $

formoterol (*Foradil*, ♣*Oxeze Turbuhaler*): 1 puff bid. Not for acute bronchospasm. Use only in combination with corticosteroids. [Trade: DPI 12 mcg, 12 & 60 blisters/pack. Canada only (Oxeze): DPI 6 & 12 mcg 60 blisters/pack.] ▶L ♀C ▶? $$$

levalbuterol (*Xopenex, Xopenex HFA*): MDI 2 puffs q4-6h prn. 0.63-1.25 mg nebulized q6-8h. 6-11 yo: 0.31 mg nebulized tid. [Trade: MDI 45 mcg/actuation, 15g 200/canister. "HFA" inhalers use hydrofluoroalkane propellant. Soln for inhalation 0.31, 0.63, 1.25 mg in 3 mL and 1.25mg in 0.5 mL unit-dose vials.] ▶L ♀C ▶? $$$

metaproterenol (*Alupent*, ♣*orciprenaline*): MDI 2-3 puffs q3-4h: 0.2-0.3 mL 5% soln nebulized q4h. 20 mg PO tid-qid >9 yo, 10 mg PO tid-qid if 6-9 yo, 1.3-2.6 mg/kg/day divided tid-qid if 2-5 yo. [Trade only: MDI 0.65 mg/actuation, 14g-200/canister. Generic/Trade: Soln for inhalation 0.4% & 0.6% in 2.5 mL unit-dose vials. Generic only: Syrup 10 mg/5 mL, Tabs 10 & 20 mg.] ▶L ♀C ▶? $$

pirbuterol (*Maxair, Maxair Autohaler*): MDI 1-2 puffs q4-6h. [Trade only: MDI 0.2 mg/actuation, 14g 400/canister.] ▶L ♀C ▶? $$$

salmeterol (*Serevent Diskus*): 1 puff bid. Not for acute bronchospasm. Use only in combination with corticosteroids. [Trade only: DPI (Diskus): 50 mcg, 60 blisters.] ▶L ♀C ▶? $$$$

terbutaline (*Brethine*, ♣*Bricanyl Turbuhaler*): 2.5-5 mg PO q6h while awake in >12 yo. 0.25 mg SC. [Generic/Trade: Tabs 2.5 & 5 mg (Brethine scored). Canada only (Bricanyl): DPI 0.5 mg/actuation, 200 per DPI.] ▶L ♀B ▶- $$

Combinations
Advair (fluticasone+salmeterol, *Advair HFA*): Asthma DPI 1 puff bid (all strengths). MDI: 2 puffs bid (all strengths). COPD with chronic bronchitis: DPI: 1 puff bid (250 /50 only). [Trade only: DPI: 100/50, 250/50, 500/50 mcg fluticasone/salmeterol per actuation; 60 doses per DPI. MDI: 45/21, 115/21, 230/21 mcg fluticasone/salmeterol per actuation; 120 doses/canister.] ▶L ♀C ▶? $$$$

Combivent (albuterol + ipratropium): 2 puffs qid, max 12 puffs/day. Contraindicated with soy or peanut allergy. [Trade only: MDI: 90 mcg albuterol/18 mcg ipratropium per actuation, 200/canister.] ▶L ♀C ▶? $$$

DuoNeb (albuterol + ipratropium, ♣*Combivent inhalation solution*): One unit dose vial. [Trade only: Unit dose: 2.5 mg albuterol/0.5 mg ipratropium per 3 mL vial, premixed. 30 & 60 vials/carton.] ▶L ♀C ▶? $$$$$

Symbicort (budesonide + formoterol): Asthma: 2 puffs bid (both strengths). [Trade only: MDI: 80/4.5, 160/4.5 mcg budesonide/formoterol per actuation; 120 doses/canister.] ▶L ♀C ▶? $$$$

INHALED STEROIDS: ESTIMATED COMPARATIVE DAILY DOSES*

Drug	Form	ADULT			CHILD (≤12 yo)		
		Low	Medium	High	Low	Medium	High
beclomethasone MDI	40 mcg/puff	2-6	6-12	>12	2-4	4-8	>8
	80 mcg/puff	1-3	3-6	>6	1-2	2-4	>4
budesonide DPI	200 mcg/dose	1-3	3-6	>6	1-2	2-4	>4
	Soln for nebs	-	-	-	0.5 mg	1 mg	2 mg
flunisolide MDI	250 mcg/puff	2-4	4-8	>8	2-3	4-5	>5
fluticasone MDI	44 mcg/puff	2-6	6-15	>15	2-4	4-10	>10
	110 mcg/puff	1-2	3-6	>6	1	1-4	>4
	220 mcg/puff	1	2-3	>3	n/a	1-2	>2
fluticasone DPI	50 mcg/dose	2-6	6-12	>12	2-4	4-8	>8
	100 mcg/dose	1-3	3-6	>6	1-2	2-4	>4
	250 mcg/dose	1	2	>2	n/a	1-2	>2
triamcinolone MDI	100 mcg/puff	4-10	10-20	>20	4-8	8-12	>12

*MDI=metered dose inhaler. DPI=dry powder inhaler. All doses in puffs (MDI) or inhalations (DPI). Reference: http://www.nhlbi.nih.gov/guidelines/asthma/execsumm.pdf

Inhaled Steroids (See Endocrine-Corticosteroids when oral steroids necessary.)

beclomethasone - inhaled (***QVAR***): 1-4 puffs bid (40 mcg). 1-2 puffs bid (80 mcg). [Trade only: MDI (non-CFC): 40 mcg & 80 mcg/actuation, 7.3g-100 actuations/canister.] ▶L ♀C ▶? $$$

budesonide - inhaled (***Pulmicort Turbuhaler, Pulmicort Respules***): 1-2 puffs daily-bid. [Trade only: DPI: 200 mcg powder/actuation, 200/canister. Respules: 0.25 mg/2 mL & 0.5 mg/2 mL unit dose.] ▶L ♀B ▶? $$$$

flunisolide - inhaled (***Aerobid, Aerobid-M, Aerospan***): 2-4 puffs bid. [Trade only: MDI: 250 mcg/actuation, 100/canister. AeroBid-M: menthol flavor. Aerospan (HFA) MDI: 80 mcg/actuation, 60 & 120/canister.] ▶L ♀C ▶? $$$

fluticasone - inhaled (***Flovent, Flovent HFA, Flovent Rotadisk***): MDI: 2-4 puffs bid. [Trade only: MDI: 44, 110, 220 mcg/actuation in 60 & 120/canister. HFA MDI: 44, 110, 220 mcg/actuation in 120/canister. DPI: 50, 100, 250 mcg/actuation delivering 44, 88, 220 mcg respectively.] ▶L ♀C ▶? $$$$

mometasone - inhaled (***Asmanex Twisthaler***): 1-2 puffs q pm or 1 puff bid. If prior oral corticosteroid therapy: 2 puffs bid. [Trade only: DPI: 220 mcg/actuation, 30, 60 & 120/canister.] ▶L ♀C ▶? $$$

triamcinolone - inhaled (***Azmacort***): 2 puffs tid-qid or 4 puffs bid; max dose 16 puffs/day. [Trade only: MDI: 100 mcg/actuation, 240/canister. Built-in spacer.] ▶L ♀D ▶? $$$$

WHAT COLOR IS WHAT INHALER? (Body then cap - Generics may differ)

Advair	purple	Combivent	clear/orange	Pulmicort	white/brown
Advair HFA	purple/light purple	Flovent HFA	orange/peach	QVAR 40 mcg	beige/grey
		Foradil	grey/beige	QVAR 80 mcg	mauve/grey
Aerobid	grey/purple	Intal	white/blue	Serevent Diskus	green
Aerobid-M	grey/green	Maxair	white/white	Spiriva	grey
Alupent	clear/blue	Maxair Autohaler	white/white	Tilade	white/white
Asmanex	pink/white			Ventolin HFA	light blue/navy
Atrovent HFA	clear/green	ProAir HFA	red/white	Xopenex HFA	blue/red
Azmacort	white/white	Proventil HFA	yellow/orange		

Leukotriene Inhibitors

montelukast (***Singulair***): Adults: 10 mg PO daily. Children 6-14 yo: 5 mg PO daily. 2-5 yo: 4 mg PO daily. 12-23 months (asthma): 4 mg (oral granules) PO daily. 6-23 months (allergic rhinitis): 4 mg (oral granules) PO daily. [Trade only: Tabs 10 mg. Oral granules 4 mg packet, 30/box. Chewable tabs (cherry flavored) 4 & 5 mg.] ▶L ♀B ▶? $$$$

zafirlukast (***Accolate***): 20 mg PO bid. Peds 5-11 yo, 10 mg PO bid. Take 1h ac or 2h pc. Potentiates warfarin & theophylline. [Trade: Tabs 10, 20 mg.] ▶L ♀B ▶- $$$

zileuton (***Zyflo***): 600 mg PO tid. Hepatotoxicity, potentiates warfarin, theophylline, & propranolol. [Trade only: Tabs 600 mg.] ▶L ♀C ▶? $$$$$

Other Pulmonary Medications

acetylcysteine (***Mucomyst***, ✦***Parvolex***): Mucolytic: 3-5 mL of 20% or 6-10 mL of 10% soln nebulized tid-qid. [Generic/Trade: Soln for inhalation 10 & 20% in 4,10 & 30 mL vials.] ▶L ♀B ▶?

aminophylline (✦***Phyllocontin***): Acute asthma: loading dose: 6 mg/kg IV over 20-30 min. Maintenance 0.5-0.7 mg/kg/h IV. [Generic: Tabs 100 & 200 mg. Oral liquid 105 mg/5 mL. Canada Trade only: Tabs controlled release (12hr) 225, 350 mg, scored.] ▶L ♀C ▶? $

cromolyn sodium (***Intal, Gastrocrom***, ✦***Nalcrom***): Asthma: 2-4 puffs qid or 20 mg nebs qid. Prevention of exercise-induced bronchospasm: 2 puffs 10-15 min prior to exercise. Mastocytosis: Oral concentrate 200 mg PO qid for adults, 100 mg qid in children 2-12 yo. [Trade only: MDI 800 mcg/actuation, 112 & 200/canister. Oral concentrate 100 mg/5 mL in 8 amps/foil pouch (Gastrocrom). Generic/Trade: Soln for nebs: 20 mg/2 mL.] ▶LK ♀B ▶? $$$

dexamethasone (***Decadron***, ✦***Maxidex***): BPD preterm infants: 0.5 mg/kg PO/IV divided q12h x 3 days, then taper. Croup: 0.6 mg/kg PO or IM x 1. Acute asthma: >2 yo: 0.6 mg/kg to max 16 mg PO daily x 2 days. [Trade: Tabs 0.5, 0.75 mg. Generic only: Tabs 0.25, 1, 1.5, 2, 4, 6 mg, various scored. Elixir 0.5 mg/5 mL. Oral solution: 0.5 mg/5 mL & 0.5 mg/0.5 mL.] ▶L ♀C ▶- $

dornase alfa (***Pulmozyme***): Cystic fibrosis: 2.5 mg nebulized daily-bid. [Trade only: soln for inhalation: 1 mg/mL in 2.5 mL vials.] ▶L ♀B ▶? $$$$$

epinephrine racemic (***S-2, ✦Vaponefrin***): Severe croup: 0.05 mL/kg/dose diluted to 3 mL w/NS. Max dose 0.5 mL. [Trade only: soln for inhalation: 2.25% epinephrine in 15 & 30 mL.] ▶Plasma ♀C ▶- $

ipratropium (***Atrovent, Atrovent HFA***): 2 puffs qid, or one 500 mcg vial neb tid-qid. Contraindicated with soy or peanut allergy (Atrovent MDI only). [Trade only: Atrovent MDI: 18mcg/actuation, 200/canister. Atrovent HFA MDI: 17 mcg/actuation, 200/canister. Generic/Trade: Soln for nebulization: 0.02% (500 mcg/vial) in unit dose vials.] ▶Lung ♀B ▶? $$

ketotifen (✦***Zaditen***): Canada only. 6 mo to 3 yo: 0.05 mg/kg PO bid. Children >3 yo: 1 mg PO bid. [Generic/Trade: Tabs 1 mg. Syrup 1mg/5 mL.] ▶L ♀C ▶- $$

nedocromil - inhaled (***Tilade***): 2 puffs qid. Reduce as tolerated [Trade only: MDI: 1.75 mg/actuation, 112/canister.] ▶L ♀B ▶? $$$

omalizumab (***Xolair***): Moderate to severe asthma with perennial allergy: 150-375 mg SC q 2-4 weeks, based on pretreatment serum total IgE level & body weight. ▶Plasma, L ♀B ▶? $$$$$

theophylline (***Elixophyllin, Uniphyl, Theo-24, T-Phyl***, ✦***Theo-Dur, Theolair***): 5-13 mg/kg/day PO in divided doses. Max dose 900 mg/day. Peds dosing variable. [Generic/Trade: Elixir 80 mg/15 mL. Trade only: Caps - Theo-24: 100, 200, 300,

400 mg. T-Phyl - 12 Hr SR tabs 200 mg. Theolair - tabs 125, 250 mg. Generic: 12 Hr tabs 100, 200, 300, 450 mg, 12 Hr caps 125, 200, 300 mg.] ▶L ♀C ▶+ $

tiotropium (**Spiriva**): COPD maintenance: Handihaler: 18 mcg inhaled daily. [Trade only: Capsule for oral inhalation, 18mcg. To be used with "Handihaler" device only. Packages of 6 or 30 capsules with Handihaler device.] ▶K ♀C ▶- $$$$

TOXICOLOGY

acetylcysteine (**Mucomyst, Acetadote, ✚Parvolex**): Contrast nephropathy prophylaxis: 600 mg PO bid on the day before and on the day of contrast. Acetaminophen toxicity: Mucomyst – loading dose 140 mg/kg PO or NG, then 70 mg/kg q4h x 17 doses. May be mixed in water or soft drink diluted to a 5% solution. Acetadote (IV) – loading dose 150 mg/kg in 200 mL of D5W infused over 60 min; maintenance dose 50 mg/kg in 500 mL of D5W infused over 4 hours followed by 100 mg/kg in 1000 mL of D5W infused over 16 hours. [Generic/Trade: solution 10%, 20%. Trade: IV (Acetadote)] ▶L ♀B ▶? $$$$

atropine (**AtroPen**): Injector pens for insecticide or nerve agent poisoning. [Trade only: Prefilled auto-injector pen: 0.25 (yellow), 0.5 (blue), 1 (dark red), 2 mg (green).] ▶LK ♀C ▶- ?

charcoal (**activated charcoal, Actidose-Aqua, CharcoAid, EZ-Char, ✚Charcodote**): 25-100 g (1-2 g/kg or 10 times the amount of poison ingested) PO or NG as soon as possible. May repeat q1-4h prn at doses equivalent to 12.5 g/hr. When sorbitol is coadministered, use only with the first dose if repeated doses are to be given. [OTC/Generic/Trade: Powder 15,30,40,120,240 g. Solution 12.5 g/60 mL, 15 g/75 mL, 15g/120 mL, 25 g/120 mL, 30 g/120 mL, 50 g/240 mL. Suspension 15g/120 mL, 25g/120ml, 30g/150 mL, 50g/240 mL. Granules 15g/120 mL.] ▶Not absorbed ♀+ ▶+ $

Cyanide Antidote Kit (amyl nitrite + sodium nitrite + sodium thiosulfate): Induce methemoglobinemia with inhaled amyl nitrite 0.3 mL followed by sodium nitrite 300 mg IV over 2-4 minutes. Then administer sodium thiosulfate 12.5 g IV. [Package contains amyl nitrite inhalant (0.3mL), sodium nitrite (300 mg/10 mL), sodium thiosulfate (12.5 g/50 mL).] ▶? ♀- ▶? $$$$$

deferasirox (**Exjade**): Chronic iron overload: 20 mg/kg PO daily; adjust dose q3-6 months based on ferritin trends. Max 30 mg/kg/day. [Trade only: Tabs for dissolving into oral suspension 125, 250, 500 mg.] ▶L ♀B ▶? $$$$$

deferoxamine (**Desferal**): Chronic iron overload: 500-1000 mg IM daily and 2 g IV infusion (≤15 mg/kg/hr) with each unit of blood or 1-2 g SC daily (20-40 mg/kg/day) over 8-24 h via continuous infusion pump. Acute iron toxicity: IV infusion up to 15 mg/kg/hr (consult poison center). ▶K ♀C ▶? $$$$$

dimercaprol (**BAL in oil**): Specialized dosing for arsenic, mercury, gold, lead toxicity; consult poison center. ▶KL ♀C ▶? $$$$$

edetate (**EDTA, Endrate, Meritate**): Specialized dosing for lead toxicity; consult poison center. ▶K ♀C ▶- $$$

ethanol (**alcohol**): Specialized dosing for methanol, ethylene glycol toxicity if fomepizole is unavailable or delayed. ▶L ♀D ▶+ $

flumazenil (**Romazicon, ✚Anexate**): Benzodiazepine sedation reversal: 0.2 mg IV over 15 sec, then 0.2 mg q1 min prn up to 1 mg total dose. Overdose reversal: 0.2 mg IV over 30 sec, then 0.3-0.5 mg q30 sec prn up to 3 mg total dose. Contraindicated in mixed drug OD or chronic benzodiazepine use. ▶LK ♀C ▶? $$$

fomepizole (**Antizol**): Specialized dosing in ethylene glycol or methanol toxicity. ▶L ♀C ▶? $$$$$

ANTIDOTES

Toxin	Antidote/Treatment	Toxin	Antidote/Treatment
acetaminophen	N-acetylcysteine	ethylene glycol	fomepizole
antidepressants*	bicarbonate	heparin	protamine
arsenic, mercury	dimercaprol (BAL)	iron	deferoxamine
benzodiazepine	flumazenil	lead	EDTA, succimer
beta blockers	glucagon	methanol	fomepizole
calcium channel	calcium chloride,	methemoglobin	methylene blue
blockers	glucagon	opioids	naloxone
cyanide	Lilly cyanide kit	organophosphates	atropine+pralidoxime
digoxin	dig immune Fab	warfarin	vitamin K, FFP

cyclic

ipecac syrup: Emesis: 30 mL PO for adults, 15 mL if 1-12 yo. [Generic/OTC: syrup.] ▶Gut ♀C ▶? $

methylene blue (*Urolene blue*): Methemoglobinemia: 1-2 mg/kg IV over 5 min. ▶K ♀C ▶? $$

penicillamine (*Cuprimine, Depen*): Specialized dosing for copper toxicity. [Trade only: Caps 125, 250 mg; tabs 250 mg.] ▶K ♀D ▶- $$$$

pralidoxime (*Protopam, 2-PAM*): Organophosphate poisoning; consult poison center: 1-2 g IV infusion over 15-30 min or slow IV injection ≥5 min (max rate 200 mg/min). May repeat dose after 1 h if muscle weakness persists. Peds: 20-50 mg/kg/dose IV over 15-30 min. ▶K ♀C ▶? $$$

succimer (*Chemet*): Lead toxicity in children ≥1 yo: Start 10 mg/kg PO or 350 mg/m² q8h x 5 days, then reduce the frequency to q12h x 2 weeks. [Trade only: Caps 100 mg.] ▶K ♀C ▶? $$$$$

UROLOGY

Benign Prostatic Hyperplasia

alfuzosin (*UroXatral*, ✱*Xatral*): BPH: 10 mg PO daily after a meal. [Trade only: extended-release tab 10 mg.] ▶KL ♀B ▶- $$$

doxazosin (*Cardura, Cardura XL*): Immediate release: Start 1 mg PO qhs, max 8 mg/day. Extended release: 4 mg PO qam with breakfast, max 8 mg/day. [Generic/Trade: Immediate-release tabs 1,2,4,8 mg. Trade: XL tabs 4, 8 mg.] ▶LK ♀C ▶? $$

dutasteride (*Avodart*): BPH: 0.5 mg PO daily. [Trade: Cap 0.5 mg.] ▶L ♀X ▶- $$$

finasteride (*Proscar*): 5 mg PO daily alone or in combination with doxazosin to reduce the risk of symptomatic progression of BPH. [Generic/Trade: Tab 5 mg.] ▶L ♀X ▶- $$$$

tamsulosin (*Flomax*): 0.4 mg PO daily, 30 min after a meal. Maximum 0.8 mg/day. [Trade only: Cap 0.4 mg.] ▶LK ♀B ▶- $$$

terazosin (*Hytrin*): Start 1 mg PO qhs, usual effective dose 10 mg/day, max 20 mg/day. [Generic/Trade: Caps 1, 2, 5, 10 mg.] ▶LK ♀C ▶? $

Bladder Agents - Anticholinergics & Combinations

darifenacin (*Enablex*): Overactive bladder with symptoms of urinary urgency, frequency and urge incontinence: 7.5 mg PO daily. May increase to max dose 15 mg PO daily in 2 weeks. Max dose 7.5mg PO daily with moderate liver impairment or when coadministered with potent CYP3A4 inhibitors (ketoconazole, itraconazole, ritonavir, nelfinavir, clarithromycin & nefazodone). [Trade only: Extended-release tabs 7.5, 15 mg.] ▶LK ♀C ▶- $$$$

flavoxate (**Urispas**): 100-200 mg PO tid-qid. [Trade: Tab 100 mg.] ▶K ♀B ▶? $$$$

hyoscyamine (**Anaspaz, A-spaz, Cystospaz, ED Spaz, Hyosol, Hyospaz, Levbid, Levsin, Levsinex, Medispaz, NuLev, Spacol, Spasdel, Symax**): 0.125-0.25 mg PO/SL q4h or prn. Extended release: 0.375-0.75 mg PO q12h. Max 1.5 mg/day. [Generic/Trade: Tab 0.125. Sublingual Tab 0.125 mg. Extended release Tab 0.375 mg. Extended release Cap 0.375 mg. Elixir 0.125 mg/ 5 mL. Drops 0.125 mg/1 mL. Trade: Tab 0.15 mg (Hyospaz, Cystospaz). Tab, orally disintegrating 0.125 (NuLev).] ▶LK ♀C ▶- $

oxybutynin (**Ditropan, Ditropan XL, Oxytrol, ♣Oxybutyn**): Bladder instability: Ditropan: 2.5-5 mg PO tid-qid, max 5 mg PO qid. Ditropan XL: 5-10 mg PO daily, increase 5 mg/day q week to 30 mg/day. Oxytrol: 1 patch twice weekly on abdomen, hips or buttocks. [Generic/Trade: Tab 5 mg. Syrup 5 mg/5 mL. Trade: Extended release tabs (Ditropan XL) 5, 10, 15 mg. Transdermal (Oxytrol) 3.9 mg/day.] ▶LK ♀B ▶? $

Prosed/DS (methenamine + phenyl salicylate + methylene blue + benzoic acid + hyoscyamine): 1 tab PO qid with liberal fluids. May turn urine/contact lenses blue. [Trade only: Tab (methenamine 81.6 mg/phenyl salicylate 36.2 mg/methylene blue 10.8 mg/benzoic acid 9.0 mg/hyoscyamine sulfate 0.12 mg). Prosed EC = enteric coated form.] ▶KL ♀C ▶? $$$

solifenacin (**VESIcare**): Overactive bladder with symptoms of urinary urgency, frequency or urge incontinence: 5 mg PO daily. Max dose: 10 mg daily (5 mg daily if CrCl<30 mL/min, moderate hepatic impairment, or concurrent ketoconazole or other potent CYP3A4 inhibitors). [Trade only: Tabs 5,10 mg] ▶LK ♀C ▶- $$$$

tolterodine (**Detrol, Detrol LA, ♣Unidet**): Overactive bladder: 1-2 mg PO bid (Detrol) or 2-4 mg PO daily (Detrol LA). [Trade only: Tabs 1, 2 mg. Caps, extended release 2, 4 mg.] ▶L ♀C ▶- $$$

trospium (**Sanctura**): Overactive bladder with urge incontinence: 20 mg PO bid. If CrCl <30 mL/min: 20 mg PO qhs. If ≥75 yo may taper down to 20 mg daily. [Trade only: Tab 20 mg] ▶LK ♀C ▶? $$$

Urised (methenamine + phenyl salicylate + atropine + hyoscyamine + benzoic acid + methylene blue, Usept): Dysuria: 2 tabs PO qid. May turn urine/contact lenses blue, don't use with sulfa. [Trade only: Tab (methenamine 40.8 mg/phenyl salicylate 18.1 mg/atropine 0.03 mg/hyoscyamine 0.03 mg/4.5 mg benzoic acid/5.4 mg methylene blue).] ▶K ♀C ▶? $$$$

UTA (methenamine + sodium phosphate + phenyl salicylate + methylene blue + hyoscyamine): 1 cap PO qid with liberal fluids. [Trade only: Cap (methenamine 120 mg/sodium phosphate 40.8 mg/phenyl salicylate 36 mg/methylene blue 10 mg/hyoscyamine 0.12 mg).] ▶KL ♀C ▶? $$$

Bladder Agents - Other

bethanechol (**Urecholine, Duvoid, ♣Myotonachol**): Urinary retention: 10-50 mg PO tid-qid. [Generic/Trade: Tabs 5, 10, 25, 50 mg.] ▶L ♀C ▶? $

desmopressin (**DDAVP, ♣Minirin**): Enuresis: 10-40 mcg intranasally qhs or 0.2-0.6 mg PO qhs. Not for children <6 yo. [Generic/Trade: Tabs 0.1, 0.2 mg; Nasal solution 0.1 mg/mL (10 mcg/ spray). Trade only: DDAVP Rhinal Tube: 2.5 mL bottle with 2 flexible plastic tube applicators with graduation marks for dosing.] ▶LK ♀B ▶? $$$$

imipramine (**Tofranil, Tofranil PM**): Enuresis: 25-75 mg PO qhs. [Generic/Trade: Tabs 10, 25, 50 mg. Trade only: Caps (Tofranil-PM) 75, 100, 125, 150 mg.] ▶L ♀B ▶? $

methylene blue (*Methblue 65, Urolene blue*): Dysuria: 65-130 mg PO tid after meals with liberal water. May turn urine/contact lenses blue. [Trade only: Tab 65 mg.] ▶Gut/K ♀C ▶? $

pentosan (*Elmiron*): Interstitial cystitis: 100 mg PO tid. [Trade only: Caps 100 mg.] ▶LK ♀B ▶? $$$$

phenazopyridine (*Pyridium, Azo-Standard, Urogesic, Prodium, Pyridiate, Urodol, Baridium, UTI Relief, ✦Phenazo*): Dysuria: 200 mg PO tid x 2 days. May turn urine/contact lenses orange. [OTC Generic/Trade: Tabs 95, 97.2 mg. Rx Generic/Trade: Tabs 100, 200 mg.] ▶K ♀B ▶? $

Erectile Dysfunction

alprostadil (*Muse, Caverject, Caverject Impulse, Edex, ✦Prostin VR*): 1 intraurethral pellet (Muse) or intracavernosal injection (Caverject, Edex) at lowest dose that will produce erection. Onset of effect is 5-20 minutes. [Trade only: Syringe system (Edex) 10, 20, 40 mcg. (Caverject) 5, 10, 20 mcg. (Caverject Impulse) 10, 20 mcg. Pellet (Muse) 125, 250, 500, 1000 mcg. Intracorporeal injection of locally-compounded combination agents (many variations): "Bi-mix" can be 30 mg/mL papaverine + 0.5 to 1 mg/mL phentolamine, or 30 mg/mL papaverine + 20 mcg/mL alprostadil in 10 mL vials. "Tri-mix" can be 30 mg/mL papaverine + 1 mg/mL phentolamine + 10 mcg/mL alprostadil in 5, 10 or 20 mL vials.] ▶L ♀- ▶- $$$$

sildenafil (*Viagra, Revatio*): Erectile dysfunction: Start 50 mg PO 0.5-4 h prior to intercourse. Max 1 dose/day. Usual effective range 25-100 mg. Start at 25 mg if >65 yo or liver/renal impairment. Pulmonary hypertension: 20 mg PO tid. Contraindicated with nitrates. [Trade only (Viagra): Tabs 25, 50, 100 mg. Unscored tab but can be cut in half. Revatio: Tabs 20 mg.] ▶LK ♀B ▶- $$$

tadalafil (*Cialis*): Start 10 mg PO ≥30-45 min prior to sexual activity. May increase to 20 mg or decrease to 5 mg prn. Max 1 dose/day. Start 5 mg (max 1 dose/day) if CrCl 31-50 mL/min. Max 5 mg/day if CrCl <30 mL/min on dialysis. Max 10 mg/day if mild to moderate hepatic impairment; avoid in severe hepatic impairment. Max 10 mg once in 72 hours if concurrent potent CYP3A4 inhibitors. Contraindicated with nitrates & alpha-blockers (except tamsulosin 0.4 mg daily). Not FDA approved for women. [Trade only: Tabs 5, 10, 20 mg.] ▶L ♀B ▶- $$$$

vardenafil (*Levitra*): Start 10 mg PO 1 h before sexual activity. Usual effective dose range 5-20 mg. Max 1 dose/day. Use lower dose (5 mg) if ≥65 yo or moderate hepatic impairment (max 10 mg). Contraindicated with nitrates and alpha-blockers. Not FDA-approved in women. [Trade: Tabs 2.5, 5, 10, 20 mg] ▶LK ♀B ▶- $$$

yohimbine (*Yocon, Yohimex*): Erectile dysfunction (not FDA approved): 5.4 mg PO tid. [Generic/Trade: Tab 5.4 mg.] ▶L ♀- ▶- $

Nephrolithiasis

acetohydroxamic acid (*Lithostat*): Chronic UTI adjunctive therapy: 250 mg PO tid-qid. [Trade only: Tab 250 mg.] ▶K ♀X ▶? $$$

citrate (*Polycitra-K, Urocit-K, Bicitra, Oracit, Polycitra, Polycitra-LC*): Urinary alkalinization: 1 packet in water/juice PO tid-qid. [Trade only: Polycitra-K packet (potassium citrate): 3300 mg. Oracit oral solution: 5 mL = sodium citrate 490 mg. Generic/Trade: Urocit-K wax (potassium citrate) Tabs 5, 10 mEq. Generic only: Polycitra-K oral solution (5 mL = potassium citrate 1100 mg), Bicitra oral solution (5 mL = sodium citrate 500 mg), Polycitra-LC oral solution (5 mL = potassium citrate 550 mg/sodium citrate 500 mg), Polycitra oral syrup (5 mL = potassium citrate 550 mg/sodium citrate 550 mg).] ▶K ♀C ▶? $$$

Index

To facilitate speed of use, index entries are shown both with page number and the approximate position on the specified page, ie, "t" is top, "m" is middle, "b" is bottom. Although the PDA software edition of the *Tarascon Pocket Pharmacopoeia* contains more than 6,000 drugs and drug names, it is physically impossible to include all of this information in pocket-sized manuals. Accordingly, a number of rarely used or highly specialized drugs appear only in our PDA edition (noted as "**PDA**"), or just in the PDA and in our larger Deluxe print edition (noted as "**D**").

133
Index

t = top of page
m = middle of page
b = bottom of page
D, PDA see page 129

137
Index

t = top of page
m = middle of page
b = bottom of page
D, PDA see page 129

138
Index

t = top of page
m = middle of page
b = bottom of page
D, PDA see page 129

139
Index

t = top of page
m = middle of page
b = bottom of page
D, PDA see page 129

**140
Index**

t = top of page
m = middle of page
b = bottom of page
D, PDA see page 129

**141
Index**

t = top of page
m = middle of page
b = bottom of page
D, PDA see page 129

143
Index

t = top of page
m = middle of page
b = bottom of page
D, PDA see page 129

Menopur D
Menostar 103t
Mentax 54b
mepenzolate 81b
meperidine 16t
mephentermine D
mephobarbital PDA
Mephyton 70t
mepivacaine 19b
Mepron 23t
mequinol 59b
mercaptopurine 108m
Meridia 121t
Meritate 125b
meropenem 28m
Merrem IV 28m
Mersyndol with Codeine 17b
mesalamine 84t
Mesasal 84t
M-Eslon 16m
mesna 108m
Mesnex 108m
Mestinon 99b
Mestinon Timespan 99b
mestranol 102
Metabolife 356 89m
Metadate 120b
Metadol
  Analgesics 16t
  Psychiatry 119m
Metaglip 62b
Metamucil 82t
Metanx D
metaproterenol 122m
metaraminol D
Metastron 108m
metaxalone 11t
metformin 62mb,65b
methacholine D
methadone
  Analgesics 16t
  Psychiatry 119m
Methadose
  Analgesics 16t
  Psychiatry 119m
methamphetamine PDA
methazolamide 111b

Methblue 65 128t
methenamine 127mb
Methergine 106m
methimazole 68b
Methitest 60t
methocarbamol 11t
methohexital 19m
methotrexate
  Analgesics 10m
  Oncology 108m
methotrimeprazine D
methoxsalen 59m
methscopolamine 84t
methsuximide D
methylcellulose 82t
methylcobalamin D
methyldopa 39t,44m
methylene blue
  Toxicology 126t
  Urology 127m,128t
methylergonovine 106m
Methylin 120b
methylphenidate 120b
methylprednisolone 61m
methylsulfomethane 91t
methyltestosterone
  Endocrine 60t
  OB/GYN 104mb
Meticorten 61b
metipranolol 111m
metoclopramide 78b
metolazone 49m
Metopirone D
metoprolol 44b, 46b
Metrika A1CNow 65m
MetroCream 54m
MetroGel 54m
MetroGel-Vaginal 107t
MetroLotion 54m
metronidazole
  Antimicr 23m, 36t
  Gastroenterol 80b

metronidazole - topical 54m
metronidazole - vaginal 107t
Metvixia 59m
metyrapone D
metyrosine D
Mevacor 42m
mexiletine 40m
Mexitil 40m
Miacalcin 71t
micafungin 21m
Micardis 38m
Micardis HCT 44b
Micardis Plus 44b
Micatin 55m
miconazole
  Dermat 55m, 59b
  OB/GYN 107t
Micort-HC Lipocream D
Micozole 107t
MICRhoGAM 107b
Microgestin Fe 102
Micro-K 67b
Micronase 65t
Micronor 102
Microzide 49m
Midamor 49t
midazolam 19m
midodrine 51t
Midrin 99t
Mifeprex 107b
mifepristone 107b
miglitol 62t
miglustat PDA
Migquin 99t
Migraine Therapy - Triptans 97b
Migra-Lief 89b
Migranal 99t
MigraSpray 89b
Migratine 99t
Migrazone 99t
mild & strong silver protein 92m
Milk of Magnesia 82m
milk thistle 91m
milrinone 51t
mineral oil 83m
Minerals 66m

Minipress 39t
Minirin
  Endocrine 71m
  Urology 128t
Minitran 50t
Minizide 44b
Minocin 35m
minocycline 35m
Minox 59m
minoxidil, topical 59m
Minoxidil for Men 59m
Mintezol 23b
Miochol-E D
Miostat D
MiraLax 82m
Mirapex 100m
Mircette 102
mirtazapine 115m
misoprostol
  Analgesics 12b
  Gastroenterol 81b
  OB/GYN 104b
mitomycin 108t
Mitomycin-C 108t
mitotane 108m
Mitotic Inhibitors 108
mitoxantrone 108t
M-M-R II 93m
Moban D
Mobic 13m
Mobicox 13m
moclobemide D
modafinil 121t
Modecate 117m
Modeten 117m
Modicon 102
Moduret 44b
Moduretic 44b
moexipril 37m, 45t
Mogadon D
molindone D
mometasone, inhaled 123m
mometasone, nasal 76b

145 Index

t = top of page
m = middle of page
b = bottom of page
D, PDA see page 129

146
Index

t = top of page
m = middle of page
b = bottom of page
D, PDA see page 129

149
Index
t = top of page
m = middle of page
b = bottom of page
D, PDA see page 129

152

Index

t = top of page
m = middle of page
b = bottom of page
D, PDA see page 129

154
Index

t = top of page
m = middle of page
b = bottom of page
D, PDA see page 129

ADULT EMERGENCY DRUGS (selected)

ALLERGY	diphenhydramine (*Benadryl*): 50 mg IV/IM. epinephrine: 0.1-0.5 mg IM (1:1000 solution), may repeat after 20 minutes. methylprednisolone (*Solu-Medrol*): 125 mg IV/IM.
HYPERTENSION	esmolol (*Brevibloc*): 500 mcg/kg IV over 1 minute, then titrate 50-200 mcg/kg/minute fenoldopam (*Corlopam*): Start 0.1 mcg/kg/min, titrate up to 1.6 mcg/kg/min labetalol (*Normodyne*): Start 20 mg slow IV, then 40-80 mg IV q10 min prn up to 300 mg total cumulative dose nitroglycerin (*Tridil*): Start 10-20 mcg/min IV infusion, then titrate prn up to 100 mcg/minute nitroprusside (*Nipride*): Start 0.3 mcg/kg/min IV infusion, then titrate prn up to 10 mcg/kg/minute
DYSRHYTHMIAS / ARREST	adenosine (*Adenocard*): PSVT (not A-fib): 6 mg rapid IV & flush, preferably through a central line or proximal IV. If no response after 1-2 minutes then 12 mg. A third dose of 12 mg may be given prn. amiodarone (*Cordarone, Pacerone*): V-fib or pulseless V-tach: 300 mg IV/IO; may repeat 150 mg just once. Life-threatening ventricular arrhythmia: Load 150 mg IV over 10 min, then 1 mg/min x 6h, then 0.5 mg/min x 18h. atropine: 0.5 mg IV, repeat prn to maximum of 3 mg. diltiazem (*Cardizem*): Rapid A-fib: bolus 0.25 mg/kg or 20 mg IV over 2 min. May repeat 0.35 mg/kg or 25 mg 15 min after 1st dose. Infusion 5-15 mg/h. epinephrine: 1 mg IV/IO q3-5 minutes for cardiac arrest. [1:10,000 solution] lidocaine (*Xylocaine*): Load 1 mg/kg IV, then 0.5 mg/kg q8-10min prn to max 3 mg/kg. Maintenance 2g in 250ml D5W (8 mg/ml) at 1-4 mg/min drip (7-30 ml/h).
PRESSORS	dobutamine (*Dobutrex*): 2-20 mcg/kg/min. 70 kg: 5 mcg/kg/min with 1 mg/mL concentration (eg, 250 mg in 250 mL D5W) = 21 mL/h. dopamine (*Intropin*): Pressor: Start at 5 mcg/kg/min, increase prn by 5-10 mcg/kg/min increments at 10 min intervals, max 50 mcg/kg/min. 70 kg: 5 mcg/kg/min with 1600 mcg/mL concentration (eg, 400 mg in 250 ml D5W) = 13 mL/h. Doses in mcg/kg/min: 2-4 = traditional renal doses, apparently ineffective) dopaminergic receptors; 5-10 = (cardiac dose) dopaminergic and beta1 receptors; >10 = dopaminergic, beta1, and alpha1 receptors. norepinephrine (*Levophed*): 4 mg in 500 ml D5W (8 mcg/ml) at 2-4 mcg/min. 22.5 ml/h = 3 mcg/min. phenylephrine (*Neo-Synephrine*): 50 mcg boluses IV. Infusion for hypotension: 20 mg in 250ml D5W (80 mcg/ml) at 40-180 mcg/min (35-160ml/h).
INTUBATION	etomidate (*Amidate*): 0.3 mg/kg IV. methohexital (*Brevital*): 1-1.5 mg/kg IV. rocuronium (*Zemuron*): 0.6-1.2 mg/kg IV. succinylcholine (*Anectine*): 1 mg/kg IV. Peds (<5 yo): 2 mg/kg IV; consider preceding with atropine 0.02 mg/kg. thiopental (*Pentothal*): 3-5 mg/kg IV.
SEIZURES	diazepam (*Valium*): 5-10 mg IV, or 0.2-0.5 mg/kg rectal gel up to 20 mg PR. fosphenytoin (*Cerebyx*): Load 15-20 "phenytoin equivalents" per kg either IM, or IV no faster than 100-150 mg/min. lorazepam (*Ativan*): 0.05-0.15 mg/kg up to 3-4 mg IV/IM. phenobarbital: 200-600 mg IV at rate ≤60 mg/min; titrate prn up to 20 mg/kg phenytoin (*Dilantin*): 15-20 mg/kg up to 1000 mg IV no faster than 50 mg/min.

CARDIAC DYSRHYTHMIA PROTOCOLS (for adults and adolescents)

Chest compressions ~100/minute. Ventilations 8-10/minute if intubated; otherwise 30:2 compression/ventilation ratio. Drugs that can be administered down ET tube (use 2-2.5 x usual dose): epinephrine, atropine, lidocaine, vasopressin, naloxone.

V-Fib, Pulseless V-Tach
Airway, oxygen, CPR until defibrillator ready
Defibrillate 360 J (old monophasic), 120-200 J (biphasic), or with AED
Resume CPR x 2 minutes (5 cycles)
Repeat defibrillation if no response
Vasopressor during CPR:
- Epinephrine 1 mg IV/IO q3-5 minutes, or
- Vasopressin 40 units IV to replace 1st or 2nd dose of epinephrine
Rhythm/pulse check every ~2 minutes
Consider antiarrhythmic during CPR:
- Amiodarone 300 mg IV/IO; may repeat 150 mg just once
- Lidocaine 1.0-1.5 mg/kg IV/IO, then repeat 0.5-0.75 mg/kg to max 3 doses or 3 mg/kg
- Magnesium 1-2 g IV/IO if suspect torsade de pointes

Asystole or Pulseless Electrical Activity (PEA)
Airway, oxygen, CPR
Vasopressor (when IV/IO access):
- Epinephrine 1 mg IV/IO q3-5 minutes, or
- Vasopressin 40 units IV/IO to replace 1st or 2nd dose of epinephrine
Consider atropine 1 mg IV/IO for asystole or slow PEA. Repeat q3-5 min up to 3 doses.
Rhythm/pulse check every ~2 minutes
Consider 6 H's: hypovolemia, hypoxia, H+ acidosis, hyper / hypokalemia, hypoglycemia, hypothermia
Consider 5 T's: Toxins, tamponade-cardiac, tension pneumothorax, thrombosis (coronary or pulmonary), trauma

Bradycardia, <60 bpm and Inadequate Perfusion
Airway, oxygen, IV
Prepare for transcutaneous pacing; don't delay if advanced heart block
Consider atropine 0.5 mg IV; may repeat to max 3 mg
Consider epinephrine (2-10 mcg/min) or dopamine (2-10 mcg/kg/min)
Prepare for transvenous pacing

Tachycardia with Pulses
Airway, oxygen, IV
If unstable and heart rate >150 bpm, then synchronized cardioversion
If stable narrow-QRS (<120 ms):
- Regular: Attempt vagal maneuvers. If no success, then adenosine 6 mg IV, then 12 mg prn up to twice.
- Irregular: Control rate with diltiazem or beta blocker (caution in CHF or pulmonary disease).
If stable wide-QRS (>120 ms):
- Regular and suspect V-tach: Amiodarone 150 mg IV over 10 min; repeat prn to max 2.2 g/24h. Prepare for elective synchronized cardioversion.
- Regular and suspect SVT with aberrancy: adenosine as per narrow-QRS above.
- Irregular and A-fib: Control rate with diltiazem or beta blocker (caution in CHF/pulmonary disease).
- Irregular and A-fib with pre-excitation (WPW): Avoid AV nodal blocking agents; consider amiodarone 150 mg IV over 10 minutes.
- Irregular and torsade de pointes: magnesium 1-2 g IV load over 5-60 minutes, then infusion.

bpm=beats per minute; CPR=cardiopulmonary resuscitation; ET=endotracheal; IO=intraosseous; J=Joules; ms=milliseconds; WPW=Wolf-Parkinson-White. Source *Circulation* 2005; 112, suppl IV.

Tarascon Publishing Order Form on Next Page

Price per Copy by Number of Copies Ordered				
Total # of each ordered →	1–9	10–49	50–99	≥100
Tarascon Pocket Pharmacopoeia				
• Classic shirt pocket edition	$11.95	$10.95	$9.95	$8.95
• Deluxe lab coat pocket edition	$19.95	$16.95	$14.95	$13.95
• PDA edition on CD, 12 month subscription	$29.95	$25.46	$23.96	$22.46
• PDA edition on CD, 3 month subscription	$8.97	$7.62	$7.18	$6.73
Other Tarascon Pocketbooks				
• *Tarascon Internal Med & Crit Care Pocketbook*	$14.95	$13.45	$11.94	$10.44
• *Tarascon Primary Care Pocketbook*	$14.95	$13.45	$11.94	$10.44
• *Tarascon Peds Emergency Pocketbook*	$14.95	$13.45	$11.94	$10.44
• *Tarascon Adult Emergency Pocketbook*	$14.95	$13.45	$11.94	$10.44
• *Tarascon Pocket Orthopaedica*	$14.95	$13.45	$11.94	$10.44
• *How to be a Truly Excellent Junior Med Student*	$9.95	$8.25	$7.45	$6.95
Tarascon Rapid Reference Cards & Magnifier				
• *Tarascon Quick P450 Enzyme Reference Card*	$1.95	$1.85	$1.75	$1.65
• *Tarascon Quick Pediatric Reference Card*	$1.95	$1.85	$1.75	$1.65
• *Tarascon Quick HTN/LDL Reference Card*	$1.95	$1.85	$1.75	$1.65
• *Tarascon Fresnel Magnifying Lens and Ruler*	$1.00	$0.89	$0.78	$0.66
Other Recommended Pocketbooks				
• *Managing Contraception*	$10.00	$9.75	$9.50	$9.25
• *OB/GYN & Infertility*	$14.95	$14.55	$14.25	$13.80
• *Airway Cam Pocket Guide to Intubation*	$14.95	$14.55	$14.25	$13.80
• *Thompson's Rheumatology Pocket Reference*	$14.95	$14.55	$14.25	$13.80
• *Reproductive Endocrinology/Infertility-Pocket*	$14.95	$14.55	$14.25	$13.80
• *Reproductive Endocrinology/Infertility-Desk*	$24.95	$24.55	$24.20	$23.80

Shipping & Handling (based on subtotal on next page order form)					
If subtotal is →	≤$12	$13-29.99	$30-75.99	$76-200	$201-700
Standard shipping	$1.00	$2.50	$6.00	$8.00	$15.00
UPS 2-day air *(no PO boxes)*	$12.00	$14.00	$16.00	$18.00	$35.00

Tarascon Pocket Pharmacopoeia® Deluxe PDA Edition

Features

- Palm OS® and Pocket PC® versions
- Meticulously peer-reviewed drug information
- Multiple drug interaction checking
- Continuous internet auto-updates
- Extended memory card support
- Multiple tables & formulas
- Complete customer privacy
  Download a **FREE 30-day trial** version at www.tarascon.com
  Subscriptions thereafter priced at **$2.29/month**

Ordering Books From Tarascon Publishing

INTERNET	**MAIL**	**FAX**	**PHONE**
Order through our OnLine store with your credit card at www.tarascon.com	Mail order & check to: **Tarascon Publishing** PO Box 517 Lompoc, CA 93438	Fax credit card orders 24 hrs/day toll free to **877.929.9926**	For phone orders or customer service, call **800.929.9926**

Name		Company name (if applicable)	
Address			
City	State	Zip	Residential ☐ Business ☐
Phone	Email		

TARASCON POCKET PHARMACOPOEIA®	Quantity	Price*
Classic Shirt-Pocket Edition		$
Deluxe Labcoat Pocket Edition		$
PDA software on CD-ROM, 3 month subscription‡		$
PDA software on CD-ROM, 12 month subscription‡		$

OTHER TARASCON POCKETBOOKS		
Tarascon Internal Medicine & Critical Care Pocketbook		$
Tarascon Primary Care Pocketbook		$
Tarascon Pediatric Emergency Pocketbook		$
Tarascon Adult Emergency Pocketbook		$
Tarascon Pocket Orthopaedica®		$
How to be a Truly Excellent Junior Medical Student		$

TARASCON RAPID REFERENCE CARDS & MAGNIFIER		
Tarascon Quick P450 Enzyme Reference Card		$
Tarascon Quick Pediatric Reference Card		$
Tarascon Quick HTN & LDL Reference Card		$
Tarascon Fresnel Magnifying Lens and Ruler		$

OTHER RECOMMENDED POCKETBOOKS		
Managing Contraception		$
OB/GYN & Infertility		$
Thompson's Rheumatology Pocket Reference		$
Airway Cam Pocket Guide to Intubation		$
Reproductive Endocrinology/Infertility Pocket edition		$
Reproductive Endocrinology/Infertility Desk edition		$

See prior page for prices / shipping. ‡Or download today at www.tarascon.com

☐ VISA ☐ Mastercard ☐ American Express ☐ Discover	*Subtotal*	$
Card number	CA only add 7.25% sales tax	$
Exp date	CID number (if available)	Shipping / handling*	$
Signature		**TOTAL**	$